Studies in Computational Intelligence

Volume 955

Series Editor

Janusz Kacprzyk, Polish Academy of Sciences, Warsaw, Poland

The series "Studies in Computational Intelligence" (SCI) publishes new developments and advances in the various areas of computational intelligence—quickly and with a high quality. The intent is to cover the theory, applications, and design methods of computational intelligence, as embedded in the fields of engineering, computer science, physics and life sciences, as well as the methodologies behind them. The series contains monographs, lecture notes and edited volumes in computational intelligence spanning the areas of neural networks, connectionist systems, genetic algorithms, evolutionary computation, artificial intelligence, cellular automata, self-organizing systems, soft computing, fuzzy systems, and hybrid intelligent systems. Of particular value to both the contributors and the readership are the short publication timeframe and the world-wide distribution, which enable both wide and rapid dissemination of research output.

Indexed by SCOPUS, DBLP, WTI Frankfurt eG, zbMATH, SCImago.

All books published in the series are submitted for consideration in Web of Science.

More information about this series at https://link.springer.com/bookseries/7092

María Eugenia Cornejo · László T. Kóczy ·
Jesús Medina-Moreno · Juan Moreno-García
Editors

Computational Intelligence and Mathematics for Tackling Complex Problems 2

Springer

Editors
María Eugenia Cornejo
Faculty of Engineering
Department of Mathematics
University of Cádiz
Cádiz, Spain

Jesús Medina-Moreno
Science Faculty
Department of Mathematics
University of Cádiz
Cádiz, Spain

László T. Kóczy
Faculty of Engineering Sciences
Széchenyi István University
Győr, Hungary

Budapest University of Technology
and Economics
Budapest, Hungary

Juan Moreno-García
Escuela de Ingeniería Industrial
Universidad de Castilla La Mancha
Toledo, Spain

ISSN 1860-949X ISSN 1860-9503 (electronic)
Studies in Computational Intelligence
ISBN 978-3-030-88816-9 ISBN 978-3-030-88817-6 (eBook)
https://doi.org/10.1007/978-3-030-88817-6

This Springer imprint is published by the registered company Springer Nature Switzerland AG
The registered company address is: Gewerbestrasse 11, 6330 Cham, Switzerland

Preface

The conjugation of computational sciences with different mathematical tools is essential in order to solve different challenges that arise in a wide-ranging knowledge areas. Nowadays, many promising research lines are being developed in this direction from the theoretical and applicational perspectives.

In this publication, computational intelligence and mathematics are combined in interesting research works that aim to give answers to complex real problems. This book collects the final versions of the highest quality papers presented at the conference 11th European Symposium on Computational Intelligence and Mathematics held on 2–5 October 2019, in Toledo (Spain). Moreover, the technical program of this conference included four excellent keynote speeches, given by Prof. José Luis Verdegay (Guidelines to solve Decision Making Problems), Prof. Joao Paulo Carvalho (Recommender Systems: Using Fuzzy Fingerprints for "Proper" Recommendations), Dr. Andreja Tepavcevic (Special lattice valued structures and approximate solutions of linear equations), and Prof. Juan Moreno-Garcia (Generating linguistic descriptions using Linguistic Petri Nets).

In the following, a summary of the papers included in this publication is presented:

Chapter "A New Exploitation Scheme in the Context of Bipolar Classifiers" presents a new exploitation approach for taking advantage of the information given by the bipolar knowledge representation paradigm. The developed approach fits a rule-based classifier in order to establish the decision rule which gives rise to the final class assignments on the basis of the bipolar scores provided by any soft classifier.

Chapter "Dynamic Maximal Covering Location Problem with Facility Types and Time Dependent Availability" focuses on location problems. In this paper, a generalization of the dynamic maximal covering location problem is proposed; this generalization includes different types of facilities where their availability change with the time.

Chapter "Implicative Linguistic Summaries" reviews some concepts associated with the usual kinds of linguistic summaries and provides the notion of extended linguistic summary with an implicative character. The notion of implicative linguistic summary and a method to compute its validity, by using fuzzy implications, are proposed.

Chapter "Comparison of Discrete Memetic Evolutionary Metaheuristics for TSP" performs a comparison among different discrete memetic evolutionary metaheuristic algorithms, such as DBMEA (Discrete Bacterial Memetic Evolutionary Algorithm) and DMTLBO (Discrete Memetic Teaching–Learning-Based Optimization), which are applicable to the optimization problem known as Traveling Salesman Problem.

Chapter "Syntactic Analysis of Sentences Using Deep Neural Networks" applies three different models of artificial neural networks to the syntactic analysis of sentences, determining the subject and the predicate of a sentence. In addition, a test to identify the kind of words that compose a sentence is shown.

A challenge in process mining enhancement is addressed in the Chapter "Estimating Remaining Time of Business Processes with Structural Attributes of the Traces", the prediction of remaining times in a business process. A new vector-based and ATS-based approach that considers structural features or attributes related to the process execution is presented. Each ATS state is characterized with a set of vectors and, based on these vectors and on the remaining times of the traces related to them, a linear regression-based predictor has been built for each state.

Chapter "Convolutional Neural Networks in the Ovarian Cancer Detection" uses artificial neural networks for the ovarian cancer detection by means of the recognition of cancer cells occurring in histopathological photos. The developed application is based on a program for pre-processing photos which was written by using Python programming language.

Chapter "Analyzing the Performance of TSP Solver Methods" analyzes the efficiency of three algorithms (Concorde, Lin-Kernighan, and DBMEA) which solve the optimization problem called Traveling Salesman Problem. As well as the tour quality and the run time properties, the proposed analysis considers the run time predictability by using polynomial, exponential, and square-root exponential models.

Chapter "A Fuzzy Model to Aggregate Performance Indicators in Sports" proposes a fuzzy model in order to identify suitable indicators for evaluating the performance of individual sports teams. The proposed fuzzy model considers the opinion of a group of sports experts and carries out consensus processes based on fuzzy aggregation functions.

Chapter "Formal Analysis of Solar Power and Weather Data" proposes a software architecture in which open data files are transformed to formal contexts, in order to use fuzzy mathematical tools for analyzing weather data and the energy produced by certain photovoltaic installations. This paper is a first approximation for improving the processes involved in the management of renewable energies and energy efficiency, by using formal concept analysis.

Chapter "A New Community Detection Problem Based on Bipolar Fuzzy Measures" proposes a modification of the Louvain algorithm in order to consider additional information provided by a fuzzy bipolar measure to detect communities. A more realistic representation and processing of human judgments, intentions, and preferences is obtained by means of this bipolar modeling.

Chapter "Novel Methods of FCM Model Reduction" is focused on the use of Fuzzy Cognitive Maps to support decision making tasks. In order to reduce the complexity of models of systems, preserving an adequate accuracy and making them simple enough to be used, the authors present two novel clustering-based Fuzzy Cognitive Map model reduction methods based on K-Means and Fuzzy C-Means clustering

Chapter "Some Implications of Interval Approach to Dimension for Network Complexity" investigates how the capabilities of feedforward networks to compute classes of regression and classification tasks are affected by the exponential growth of the number of highly uncorrelated tasks.

Chapter "Social Indexes Segregation Based on MEOWA and MOOWA Aggregation Operators" work makes use of OWA aggregator operators with the goal of constructing indexes or subindexes that allow aggregating information about variables that cannot be ranked, since all of them present the same importance.

The detection of fake news is the goal of the Chapter "Combining Conceptual Graphs and Sentiment Analysis for Fake News Detection". Making use of text mining and sentiment analysis techniques, a prototype of a knowledge-based system for the detection of fake news using reliable information sources is presented.

In Chapter "A Weakened Notion of Congruence to Reduce Concept Lattices", the problem of the attribute and size reduction of concept lattices in formal concept analysis is addressed. By means of a weaker notion of congruence relation, called weak-congruence, the authors generate a partition of the set of concepts of the concept lattice, guaranteeing that each subset of the partition forms a closed algebraic substructure and preserves the main information.

Knowledge representation by means of implications within the theory of formal concept analysis is the topic considered in the Chapter "Interactive Search by Using Minimal Generators". Making use of logic, minimal generators can be obtained from implications. In this work, a new logic-guided implication-based algorithm to infer the minimal generators of closed attribute sets is introduced.

Chapter "Knowledge Implications in Multi-adjoint Concept Lattices" introduces a contribution in the theory of formal concept analysis. An alternative definition of implication between the attributes of a context is presented. This contribution has a strong potential in tasks related to the elimination of unnecessary data.

Chapter "Some Relationships Between the Notions of f-Inclusion and f-Contradiction" shows interesting relationships between the notions of f-inclusion and f-weak contradiction, by using an involutive negation operator and isotone Galois connections. This study opens the way to a further interaction between relationships in mathematical morphology and formal concept analysis.

Chapter "Decomposition Integrals for Interval-Valued Functions" addresses the problem of decomposition integrals for interval-valued functions. In this contribution, the author presents two different extensions of this integral, one based on Aumann integral and the other one based on interval operators. It is proven that these two different approaches turn out to be equivalent and, as a consequence, both lead to the same integral.

Chapter "Relational Powerset Theories" deals with the problem of extending various structures originally defined on the set of L-valued fuzzy sets to the powerset objects defined in categories, using the general powerset theory of objects introduced by Rodabaugh.

Chapter "On Some Categories Underlying Knowledge Graphs" presents a mathematical approach to analyze and infer Resource Description Framework (RDF) based data which represent descriptions of entities. Specifically, this approach provides a categorical structure for RDF data, by constructing a category of isotone Galois connections arising from formal contexts associated with different properties of a given RDF data.

Chapter "On Two Categories of Many-Level Fuzzy Morphological Spaces" focuses on researching mathematical morphology. Many-level versions of L-fuzzy relational erosion and dilation operators are introduced in order to define two kinds of many-level fuzzy morphological spaces. A continuous transformation of such spaces giving rise to two categories of many-level fuzzy morphological spaces is also included.

Chapter "Congruences on Lattices and Lattice-Valued Functions" applies lattice-valued functions (fuzzy sets) to characterize congruences on complete lattices, making use of the connections among congruences and homomorphisms. Conversely, the authors prove that for every complete congruence on a complete lattice is the kernel of a residuated map determined by its cuts.

Chapter "Some Roughness Features of Fuzzy Sets" is devoted to study rough calculus. Following Pawlak's ideas accurately, the author investigates some roughness features of fuzzy sets, looking at the fuzzy membership functions from the point of view of rough continuity and rough approximation of real functions.

Finally, it is important to thank all those who have contributed to make possible this publication, including authors, members of the program committee, and reviewers, since without their invaluable assistance this book would not have been possible. We also acknowledge the support received from the University of Cádiz, the Hungarian Fuzzy Association, the Szechenyi Istvan University, the University of Toledo and the State Research Agency (AEI) and the European Regional Development Fund (FEDER) project TIN2016-76653-P. A word of thanks is also due to EasyChair, for the facilities provided in the submission/acceptance of the papers, and in the preparation of this book.

Cádiz, Spain — María Eugenia Cornejo
Győr, Hungary — László T. Kóczy
Cádiz, Spain — Jesús Medina-Moreno
Toledo, Spain — Juan Moreno-García

Contents

A New Exploitation Scheme in the Context of Bipolar Classifiers

Guillemo Villarino, Daniel Gómez, J. Tinguaro Rodríguez, and Alberto Fernández

Abstract From the basis of the combination of supervised classification and bipolar knowledge representation, we explore in this proposal a new approach for the exploitation of the bipolar classification scores. The proposed approach is based on fitting a rule-based classifier to determine the decision rule that produces the final class assignments on the basis of the bipolar scores provided by any soft classifier. Therefore, the proposed method can be seen as a generalization of ROC-curve based decision rules, that takes advantage of the extra information introduced through the bipolar knowledge representation. The presented experimental results shows the feasibility of the proposed approach.

Keywords Supervised classification · Bipolar knowledge representation · Soft information · Decision rule · Aggregation

Supported by FORAID Research Group by means of grant PGC2018-096509-B-I00 of the Government of Spain.

G. Villarino · D. Gómez (✉)
Facultad de Estudios Estadísticos, Universidad Complutense de Madrid, Av. Pueta de Hierro s/n, Madrid, Spain
e-mail: dagomez@estad.ucm.es

G. Villarino
e-mail: gvillari@ucm.es
URL: http://www.foraid-ucm.com/index.html

J. T. Rodríguez
Facultad de Ciencias Matemáticas, Universidad Complutense de Madrid, Plaza de las Ciencias 3, Madrid, Spain

A. Fernández
Department of Computer Science and Artificial Intelligence, CITIC-UGR, University of Granada, 18071 Granada, Spain

M. E. Cornejo et al. (eds.), *Computational Intelligence and Mathematics for Tackling Complex Problems 2*, Studies in Computational Intelligence 955,
https://doi.org/10.1007/978-3-030-88817-6_1

1 Introduction

In machine learning and supervised classification tasks, a classifier is usually understood as a function that assigns a final class to each item (i.e. a vector of values quantifying its characteristics) in a crisp way. However, most classifiers produce some kind of soft scores before to make the final decision. The most widely extended procedure to make such decision is the so-called *maximum rule*, that assigns items to the class with maximum value of the mentioned soft score. However, the application of such method is not a suitable solution in several problems, as imbalanced or highly overlapped classification [7]. An exception to the wide application of the maximum-rule occurs in the context of binary classification, in which the search and usage of optimal cut-points of the Receiving Operating Characteristic (ROC) curve has also found an extended application.

Therefore, this work is born from the idea of taking advantage of this soft scores the classifiers produce right before carrying out the final class assignment. Therefore, its main goal is to asses whether the proposed approach for avoiding the maximum-rule by managing the soft scores prior to make the final class assignment outperforms the basis results.

Section 2 contains the preliminaries regarding the bipolar knowledge representation. In Sect. 3, the proposed CART-based exploitation approach is introduced. Section 4 is devoted to present the results of a detailed experimental design. Conclusions and final remarks are shown in Sect. 5.

2 Bipolar Knowledge Representation in Supervised Classification

A methodology to introduce a bipolar knowledge representation framework in the context of supervised classification was proposed in [10]. The main idea of this proposal is to assume a certain dissimilarity structure on the set of classes that allows to obtain pairs $(ev_i^+(x), ev_i^-(x))$ of positive and negative (i.e. bipolar) scores or evidence degrees regarding the classification of an item x in each of the available classes i, $i = 1, \ldots, c$, where c denotes the number of classes. Such dissimilarity structure is determined through a dissimilarity matrix $D = (d_{ij})$, $i, j = 1, \ldots, c$, where $d_{ij} \in [0, 1]$ denotes the degree up to which class j is dissimilar to class i, and $d_{ii} = 0 \, \forall i$.

As exposed in [10], a natural way to obtain the negative evidence scores $ev_i^-(x)$ is to consider that they arise from the application of the dissimilarity structure to the scores $ev_j(x) = ev_j^+(x)$ provided by the classifier. Although these negative degrees may be built in different ways, we consider in this work the definition given in (1), where D_i denotes the ith row of the considered dissimilarity matrix D and $ev^+(x) = (ev_1^+(x), \ldots, ev_c^+(x))^t$:

$$ev_i^-(x) = \sum_{j \neq i} d_{ij} ev_j^+(x) = \sum_{j=1}^{c} d_{ij} ev_j^+(x) = D_i ev^+(x), \tag{1}$$

Once these positive and negative scores are available, different options can be considered in order to exploit them for classification purposes. In [10], an aggregation methodology was explored, and two different aggregation functions, additive and logistic, were proposed. These aggregated or adjusted scores were then exploited using the maximum-rule to provide the final class assignment.

Definition 1 Let $ev_i^+(x)$, $ev_i^-(x)$ be as before the positive and negative scores. The additive and logistic adjusted score for class C_i are defined as

$$ev_i^{add}(x) = \max\{0, ev_i^+(x) - ev_i^-(x)\}. \tag{2}$$

$$ev_i^{log}(x) = \begin{cases} 1 - e^{-\frac{ev_i^+(x)}{ev_i^-(x)}} & if\ ev_i^-(x) > 0 \\ 1 & otherwise \end{cases} \tag{3}$$

Nevertheless, the aggregation scheme is not the only way to manage the bipolar evidence and therefore a new exploitation paradigm is explored in this work to make the final class assignment based on the former pairs of positive and negative evidences for each class. Instead of considering some kind of operator to aggregate the bipolar information into a single adjusted evidence degree, applying then the maximum rule to make the final decision, in this new scheme a *general exploitation function* is used to make the final crisp classification from the basis of the positive and negative degrees as shown in Definition 2.

Definition 2 Let $ev^+ = (ev_1^+, \ldots, ev_c^+) y ev^- = (ev_1^-, \ldots, ev_c^-)$ be the positive and negative evidences provided by the base classifier and Φ a general exploitation function. Then, the assigned class for an instance x is determined by the selected *general exploitation function* Φ through its own prediction engine when the bipolar evidences for instance x, $ev^+(x) = (ev^+(x)_1, \ldots, ev_c^+(x))$ and $ev^-(x) = (ev_1^-(x), \ldots, ev_c^-(x))$ is considered as the input information.

$$Class(x) = C_i \ \ if\ and\ only\ if\ \ \Phi(ev^+(x), ev^-(x)) = C_i,\ i \in \{1, \ldots, c\} \tag{4}$$

Hereby, the final class assignment is reached by applying the *general exploitation function*, Φ, upon the positive and negative evidences.

In this frame, a particular exploitation scheme based on Rule Based Systems (RBS) merits being explored.

3 A New CART-Based Approach to Exploit Bipolar Scores

Let us now describe the proposed exploitation scheme of the bipolar scores. As it is well-known, any soft classifier returns after a training process a vector of scores $ev(x) = (ev_1(x), \ldots ev_c(x))$ quantifying the strength of association of training patterns x with each class. Typically, classifiers then exploit these scores by applying the maximum rule or selecting the optimal ROC-curve threshold in order to produce the final decision or class assignment. Instead of these options, our proposal is to first introduce the above exposed bipolar representation of the scores, and then fit a rule-based classifier such as CART [2] to these bipolar scores as decision rule. This bipolar-representation-plus-CART-based-decision mechanism is inserted into an evolutionary search of the dissimilarity matrix (see Sect. 2) in order to obtain the dissimilarity structure that produces the best results.

As introduced in Sect. 2, a general exploitation function can be used to make the final class assignment from the basis of the bipolar evidences (see Definition 2). Thus, let us now consider a meta-classifier Θ as the exploitation function in the way shown in Definition 3.

Definition 3 Let $ev^+ = (ev_1^+, \ldots, ev_c^+) y ev^- = (ev_1^-, \ldots, ev_c^-)$ be the positive and negative evidences provided by the base classifier and Θ a meta-classifier. Then, the assigned class for an instance x is determined by the crisp classification provided by the meta-classifier Θ through its own prediction engine when the bipolar evidences for instance x, $ev^+(x) = (ev^+(x)_1, \ldots, ev_c^+(x))$ and $ev^-(x) = (ev_1^-(x), \ldots, ev_c^-(x))$ is considered as input information.

$$Class(x) = C_i \;\; if \; and \; only \; if \;\; \Theta(ev^+(x), ev^-(x)) = C_i, \;\; i \in \{1, \ldots, c\} \quad (5)$$

Therefore, the proposed method works as follows: first, a base or reference soft classifier is fitted to the training data. This produces a vector $ev^+(x) = ev(x)$ of soft scores for each pattern x in the training sample. Then, an evolutionary search of the optimal dissimilarity matrix D is carried out. This search in turn proceeds as shown in Fig. 1: For each matrix D tried, the negative scores $ev^-(x)$ are obtained from the provided scores as described in Sect. 2, and then a CART classification tree (with default parameters) is fit using the bipolar pairs $(ev_i^+(x), ev_i^-(x))$, $i = 1, \ldots, c$, as input features and the training class labels as target. In this way, each training instance is assigned a class by CART, and a training performance measure can be obtained to evaluate the matrix D being tried. In this work, performance is measured using the Kappa index [5]. The genetic search then returns an optimal matrix D^*, as well as a fitted CART rule model that determines how patterns have to be classified according to the bipolar evidence obtained after applying D^* to the soft scores provided by the base classifier. These matrix D^* and CART rules can be then applied to predict new patterns or test instances after the base classifier provides the corresponding scores, hopefully enhancing its predictive power and enabling a more confident output by balancing the different concepts to be learnt.

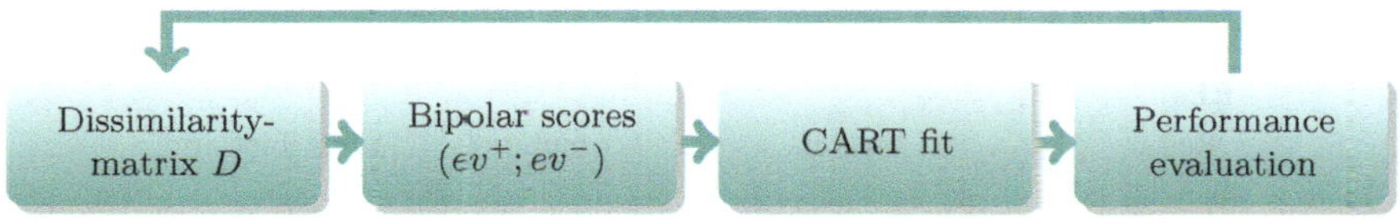

Fig. 1 Evolutionary search of the dissimilarity matrix D

The dissimilarity matrix can be found by means of some meta-heuristic. In this work a Grey Wolf Optimizer (GWO) was considered. GWO is a bio-inspired algorithm that mimics the hunting mechanism and leadership hierarchy of grey wolves [8].

Once the theoretical framework has been presented, a complete experimental study to assess the behaviour of this new bipolar-representation-plus-CART-based-decision mechanism in the context of supervised classification is shown in Sect. 4.

4 Experimental Study

This section is devoted to provide a comparison of the performance of the proposed method with that of the additive and logistic approaches introduced in [10]. To this aim, a complete experimental study in the context of three-class classification problems is presented next. Therefore, departing from the soft scores given by each base algorithm and following the exposition presented in Sects. 2 and 3, three different GWO searches are conducted, one for each evaluated method (additive and logistic aggregations, CART-based exploitation), in order to obtain the best dissimilarity structures for each method.

Regarding the base classifiers, in this work we selected two competitive classification algorithms such as Random Forest (RF) [3] and eXtreme Gradient Boosting (XGB) [4]. The three bipolar approaches together with the base classifier are fitted and tested on a benchmark of 30 multi-class datasets from the KEEL dataset repository [9], using a 5-fold cross-validation experimental framework. To transform multi-class datasets into three-class ones, we have taken as class C0 and C1 the originals closest to 20% of instances, and as class C2 the union of the remaining classes. Finally, statistical tests has been applied to rigorously assess the results. See [6] for further details of the statistical tests applied.

The objective is thus to check whether the proposed dissimilarity-driven bipolar representation model enables an improvement of the performance of the base classifiers, under different exploitation strategies and for different base classifiers.

Tables 1 and 2 show, for each dataset, the mean Kappa index of each method on the five test sets. The CART-based exploitation obtains the best results in terms of the global kappa mean in the test sets for both RF and XGB. Since the base classifiers achieve a perfect performance in many training sets, both additive and logistic aggregations had no margin to improve its performance, and a null dissimilarity structure

Table 1 RF classifier

	RF							
	Ref		bipAdd		bipLog		bipCart	
	Tr.	Tst.	Tr.	Tst.	Tr.	Tst.	Tr.	Tst.
Aba	1	0.112	1	0.112	1	0.112	1	**0.170**
Aut	1	0.698	1	0.698	1	0.698	1	0.698
BAl	0.616	0.540	0.616	0.540	0.616	0.540	0.616	0.540
Car	0.995	0.869	1	0.869	0.995	0.869	0.995	0.869
Cle	0.730	0.501	0.948	**0.506**	0.730	0.501	0.980	0.501
Con	0.788	0.269	0.814	**0.284**	0.800	0.278	0.821	0.282
Der	1	0.995	1	0.995	1	0.995	1	0.990
Eco	1	0.767	1	0.767	1	0.767	1	0.767
Fla	0.797	0.785	0.807	0.781	0.797	0.785	0.797	0.785
Gla	1	0.673	1	0.673	1	0.673	1	0.673
Hay	0.886	0.715	0.886	**0.721**	0.886	0.718	0.886	**0.721**
Iri	1	0.953	1	0.953	1	0.953	1	**0.963**
Led	0.752	0.678	0.775	0.687	0.776	0.687	0.775	**0.703**
Lin	0.990	0.681	1	0.683	0.996	0.710	0.990	**0.724**
Nty	1	0.931	1	0.931	1	0.931	1	**0.938**
Nur	0.999	0.983	1	0.983	0.999	0.983	0.999	0.983
Pag	1	0.828	1	0.828	1	0.828	1	**0.833**
Pen	1	0.906	1	0.906	1	0.906	1	0.906
Sat	1	0.907	1	0.907	1	0.907	1	0.910
Seg	1	0.993	1	0.993	1	0.993	1	0.995
Shu	1	0.996	1	0.996	1	0.996	1	0.996
Spl	1	0.598	1	0.598	1	0.598	1	**0.698**
Tae	0.953	0.467	0.954	0.467	0.953	0.467	0.953	0.467
Thy	1	0.909	1	0.909	1	0.909	1	0.909
Veh	1	0.950	1	0.950	1	0.950	1	0.950
Vow	1	0.862	1	0.862	1	0.862	1	0.862
Win	1	0.972	1	0.972	1	0.972	1	0.972
Wnq	1	0.528	1	0.528	1	0.528	1	0.528
Yea	1	0.362	1	0.362	1	0.362	1	**0.385**
Zoo	1	1	1	1	1	1	1	1
Mean	0.950	0.748	0.960	0.749	0.952	0.749	0.960	**0.757**

$(D = 0)$ is then considered. Nevertheless, the strength of our novel CART-based exploitation lies in its capability to outperform the base results in test even in this tough scenario. In fact, even considering the null dissimilarity matrix, a CART tree is fitted considering only rules regarding the positive evidence. This rule base will be

Table 2 XGB classifier

	XGB							
	Ref		bipAdd		bipLog		bipCart	
	Tr.	Tst.	Tr.	Tst.	Tr.	Tst.	Tr.	Tst.
Aba	0.919	0.134	0.958	0.152	0.925	0.177	0.919	**0.179**
Aut	1	0.734	1	0.734	1	0.734	1	**0.762**
BAl	0.605	0.569	0.612	0.577	0.605	**0.588**	0.610	0.573
Car	1	0.986	1	0.986	1	0.986	1	0.986
Cle	0.642	0.546	0.819	0.496	0.669	0.392	0.642	0.546
Con	0.415	0.308	0.456	**0.311**	0.415	0.308	0.415	0.308
Der	1	0.993	1	0.993	1	0.993	1	0.982
Eco	0.925	0.744	0.940	0.704	0.925	0.744	0.925	0.744
Fla	0.795	0.789	0.802	0.787	0.799	0.787	0.797	**0.791**
Gla	0.997	0.652	1	0.617	1	0.652	0.987	**0.673**
Hay	0.874	0.708	0.882	0.683	0.874	**0.743**	0.874	0.717
Iri	1	0.937	1	0.937	1	0.937	1	**0.943**
Led	0.755	0.667	0.767	0.675	0.755	0.676	0.770	**0.689**
Lin	0.961	0.643	0.970	0.686	0.961	**0.698**	0.961	0.688
Nty	0.999	0.913	1	0.916	1	**0.927**	1	0.913
Nur	1	0.998	1	0.997	1	0.997	1	0.998
Pag	0.974	0.805	0.992	**0.835**	0.975	0.812	0.992	0.807
Pen	1	0.915	1	0.915	1	0.915	1	**0.916**
Sat	1	0.919	1	0.919	1	0.919	1	0.918
Seg	1	0.994	1	0.994	1	0.994	1	0.995
Shu	1	0.990	1	0.990	1	0.990	1	0.990
Spl	1	0.552	1	0.552	1	0.552	1	**0.553**
Tae	0.938	0.373	0.944	0.355	0.938	0.373	0.944	**0.417**
Thy	1	0.866	1	0.866	1	0.866	1	**0.877**
Veh	1	0.954	1	0.954	1	0.954	1	0.954
Vow	1	0.903	1	0.903	1	0.903	1	0.903
Win	1	0.961	1	0.961	1	0.961	1	**0.980**
Wnq	0.999	0.504	1	**0.511**	0.999	0.506	0.999	0.506
Yea	0.632	0.364	0.663	**0.389**	0.632	0.365	0.648	0.387
Zoo	1	1	1	1	1	1	1	1
Mean	0.914	0.747	0.927	0.747	0.916	0.748	0.916	**0.757**

applied to make the final decision in the testing set, allowing to improve the reference results.

In order to add statistical support to these findings, we carried out a Friedman Aligned Rank Test for multiple comparisons along the 3 approaches together with the reference. This test shows near zero p-values for both RF and XGB classifiers,

Table 3 Average rankings of the algorithms (aligned Friedman), associated p-values and holm test APV for each algorithm

Algorithm	Rank RF	Rank XGB
Ref	71.85	72.73
BipAdd	64.90	66.48
BipLog	66.38	61.16
BipCart	**38.87**	**41.62**
Friedman p-val	0.000034	0.000029
Holm APV (Best vs. Ref)	**0.00072**	**0.00159**

Table 4 Wilcoxon test to compare the novel CART based bipolar approach (R^+) against the others (R^-)

Comparison	R^+	R^-	p-val
RFbipCart versus RFRef	358.0	107.0	**0.0094**
RFbipCart versus RFbipAdd	305.5	129.5	**0.0523**
RFbipCart versus RFbipLog	356.0	109.0	**0.0104**
XGBbipCart versus XGBRef	403.5	61.5	**0.00024**
XGBbipCart versus XGBbipAdd	315.0	120.0	**0.0238**
XGBbipCart versus XGBbipLog	310.5	154.5	0.0878

which means that there are significant differences among the results. Moreover, a post-hoc Holm test is applied for contrasting which alternatives are outperformed by the control method (i.e. our proposed CART-based exploitation) and the reference in a multi-comparison framework. Table 3 reflects the average rankings for each approach together with the Holm adjusted p-value (APV) for each reference classifier. Considering the low APV for both base algorithms, it can be concluded that the proposed method achieves a significant improvement.

Moreover, to compare the behaviour of the proposed method against the others in a pairwise scheme, a Wilcoxon test is conducted. Again, the p-values in Table 4, support the rejection of the null hypothesis of equality in results for every pair of methods considered with a confidence level of 90%. Thus, the novel approach outperforms not only the reference but also the other two bipolar classifiers. This makes the CART-based exploitation method a very suitable solution to confront high accurate base classifiers such as those considered here.

5 Concluding Remarks

In this paper, a new CART-based exploitation scheme was proposed for taking advantage of the positive and negative affects given by the bipolar knowledge representation paradigm. Several comparisons have been carried out to asses the behaviour of this novel method evidencing a significant improvement in results of this approach with respect not only to the reference soft classifiers but also to the remainder two bipolar methods, additive and logistic.

Taking into account the results obtained in the experimental study, our new exploitation model seems to be a suitable solution to handle highly accurate classifiers such as RF and XGB due to its aptness for improving the testing results when a perfect fit was reached by the reference in train.

Future researches will be devoted to extend this framework to a general n-class context, as well as to consider other rule-base classifiers such as FARC [1] for exploiting the base soft information.

References

1. Alcalá-Fdez, J., Alcalá, R., Herrera, F.: A fuzzy association rule-based classification model for high-dimensional problems with genetic rule selection and lateral tuning. IEEE Trans. Fuzzy Syst. **19**(5), 857–872 (2011)
2. Breiman, L.: Classification and Regression Trees. Kluwer Academic Publishers, New York (1984)
3. Breiman, L.: Random forests. Mach. Learn. **40**, 5–32 (2001)
4. Chen, T., Guestrin, C.: XGBoost: a scalable tree boosting system. KDD (2016)
5. Cohen, J.: A coefficient of agreement for nominal scales. Educ. Psychol. Meas. **20**, 37–46 (1960)
6. Garcia, S., Fernandez, A., Luengo, J., Herrera, F.: Advanced nonparametric tests for multiple comparisons in the design of experiments in computational intelligence and data mining: experimental analysis of power. Inf. Sci. **180**(10), 2044–2064
7. Kotsiantis, S., Kanellopoulos, D., Pintelas, P.: Handling imbalanced datasets: a review. GESTS Int. Trans. Comput. Sci. Eng. **30** (2006)
8. Mirjalili, S., Mirjalili, S.M., Lewis, A.: Grey wolf optimizer. Adv. Eng. Softw. **69**, 46–61 (2014). https://doi.org/10.1016/j.advengsoft.2013.12.007
9. Triguero, I., et al.: KEEL 3.0: an open source software for multi-stage analysis in data mining. Int. J. Comput. Intell. Syst. **10**, 1238–1249 (2017)
10. Villarino, G., Gómez, D., Rodríguez, J.T., et al.: A bipolar knowledge representation model to improve supervised fuzzy classification algorithms. Soft Comput. **22**, 5121–5146 (2018). https://doi.org/10.1007/s00500-018-3320-9

Dynamic Maximal Covering Location Problem with Facility Types and Time Dependent Availability

Cynthia Porras, Jenny Fajardo, Alejandro Rosete, Raúl E. Álvarez, and David A. Pelta

Abstract Location problems have been highly studied in the field of logistics. A classic location model is the maximal covering location problem (MCLP), which attempts to locate a limited number of facilities in order to maximize the overall covered demand. The dynamic MCLP (DMCLP) is a well-known generalization where the facilities are located in several periods of time. In this contribution we propose a generalization of the dynamic maximal covering location problem, which also includes different facility types with their availability changing over time. We present the model and we show two examples, to illustrate the amount of information that can be considered besides the coverage value attained.

Keywords Dynamic maximal covering location problem · Facility types

1 Introduction

Location problems have been successfully used in the field of logistics and distribution of services. A well-known location problem is the maximal covering location

D. Pelta acknowledges support of Project TIN2017-86647-P (Spanish Ministry of Economy and Competitiveness, includes FEDER funds from the European Union).

C. Porras · A. Rosete · R. E. Álvarez
Universidad Tecnológica de La Habana "José Antonio Echeverría" (CUJAE), 114/Ciclovia and Rotonda, Havana, Cuba
e-mail: cporrasn@ceis.cujae.edu.cu

A. Rosete
e-mail: rosete@ceis.cujae.edu.cu

J. Fajardo
Universidad de Deusto, Bilbao, Spain
e-mail: fajardo.jenny@deusto.es

D. A. Pelta (✉)
Department of Computer Science and AI, University of Granada, Granada, Spain
e-mail: dpelta@decsai.ugr.es

M. E. Cornejo et al. (eds.), *Computational Intelligence and Mathematics for Tackling Complex Problems 2*, Studies in Computational Intelligence 955,
https://doi.org/10.1007/978-3-030-88817-6_2

Table 1 Papers addressing dynamism, different facility types and time dependent availability

Paper	Dynamic	Facility types	Time-dependent availability
Schilling [11]	Yes	No	No
Moore and ReVelle [7]	No	Yes	No
Zarandi et al. [13]	Yes	No	No
Stanimirović et al. [12]	No	Yes	No
Heyns and van Vuuren [5]	No	Yes	No
Church and Murray [3]	No	Yes	No
Porras et al. [9]	Yes	Yes	No

problem (MCLP) proposed by ReVelle and Church in [2]. It attempts to locate a limited number of facilities in order to maximize the coverage over a set of demand nodes. Complex scenarios gave raise to different MCLP variants [8], from which we focus on considering heterogeneous facilities and dynamism.

One of the first models that take into account different types of facilities was proposed in [7]. Authors considered different coverage radius and hierarchy. Recently in [3], a literature review was made and a model with facility types was also presented.

Another type of model (called "dynamic") takes into account variations of the problem' parameters along the time. For example, the demand fluctuates over time. The first proposal of dynamic MCLP was published by Schilling [11]. In [13] a model which allows to locate the facilities in several time periods was proposed.

In this work, we depart from the following situation. We have a set of available WiFi hotspots that can operate with a strong, medium or weak signal (thus affecting the coverage provided). At each period, we need to decide which hotspots will be turned on and with which signal type. Several constraints make the problem harder to solve: the number of hotspots using each signal type is limited (we can not put all the hotspots with a strong signal), the total number of hotspots turned on at the end of the available periods is fixed and the type of signals available for every hotspot is period dependent.

To address this problem, we propose a MCLP model with dynamism, different facility types where their availability changes with the time.

Table 1 displays a showcase of papers addressing dynamism, different facility types and time dependent availability models. As it can be observed, none of them modeled all these features simultaneously. Moreover, dynamism was not combined with facility types, and vice versa.

In this context, the aim of this contribution is twofold: firstly, to provide a model for this new covering problem that is now named as the dynamic maximal covering location problem with facility types (DMCLP-FT) and secondly, to provide illustrative examples to analyze the distributions of the facilities and the coverage obtained by a local search algorithm. The paper is organized as follows: Section 2 describes the mathematical model of DMCLP-FT. Section 3 presents a description of examples in order to characterize the new model. Finally, Sect. 4 provides some final comments and future lines of research.

2 Dynamic Maximal Covering Location Problem with Facility Types and Time Dependent Availability

The proposed model is a generalization of DMCLP with several facility types [9], but now considering that their availability change over time. The mathematical formulation of the dynamic maximal covering location problem with facility types (DMCLP-FT) is the following:

- i, I: the index and set of demand nodes.
- j, J: the index and set of facility sites.
- t, T: the index and numbers of time periods.
- k, K: index and numbers of facility types.
- a_{it}: population or demand for the node i in period t.
- d_{ij}: the shortest distance (or time) from demand node i to the facility j.
- p_k: number of type k facilities to be located.
- S_k: coverage radius of a type k facility. It is the maximal distance (or time) between a demand node and a facility to be considered as covered.
- $W_{jtk} \in 0; 1$: a binary variable, 1 if the facility j have available the type k in period t, 0 otherwise.
- N_{itk}: $\{j|d_{ij} \leq S_k$ and $W_{jtk} = 1\}$ set of potential facility locations that can cover the demand generated in node i if a facility with type k is placed at the facility j in period t.
- X_{jtk}: $\{0, 1\}$ a binary variable, 1 if a facility with type k is placed at node j in period t, 0 otherwise.
- Y_{it}: $\{0, 1\}$ a binary variable, 1 if the demand node i in period t is covered by one or more facilities, 0 otherwise.

The objective function is:

$$Maximize : Z = \sum_{t \in T} \sum_{i \in I} a_{it} Y_{it} \tag{1}$$

Subject to:

$$Y_{it} \leq \sum_{k \in K} \sum_{j \in N_{itk}} X_{jtk}, \forall i \in I, \forall t \in T \tag{2}$$

$$\sum_{t \in T} \sum_{j \in J} X_{jtk} = p_k, \forall k \in K \tag{3}$$

$$\sum_{k \in K} p_k < J * T \tag{4}$$

$$\sum_{t \in T} \sum_{j \in J} W_{jtk} > p_k, \forall k \in K \tag{5}$$

$$\sum_{t \in T} \sum_{k \in K} W_{jtk} \geq 1, \forall j \in J \tag{6}$$

$$\sum_{k \in K} X_{jtk} \leq 1, \forall j \in J, \forall t \in T \tag{7}$$

$$\sum_{k \in K} X_{jtk} \leq W_{jtk}, \forall j \in J, \forall t \in T \tag{8}$$

The objective function (1) is aimed to maximize the coverage over the demand nodes. Constraint (2) shows that a demand node can be covered if an open facility with type k in period t belong to set Nitk. Constraint (3) shows that the number of open facilities j of type k must be equal to the value p_k. Constraint (4) shows that the total of facilities with type k that will open must be less than the total of available facilities.

Constraint (5) shows that the availability of facilities with type k must be greater than the number of facilities with type k that will be opened in all periods of time. Constraint (6) shows that a facility must have at least one type in some period t. Constraint (7) shows that the facility j can only be opened with one type. Finally, constraint (8) shows that a facility can be opened with one type k if has available that type k.

The proposed model is a generalization of the MCLP, since reducing the variables to: $T = 1$ and $K = 1$ the model becomes a standard MCLP.

Intuitively, a solution can be represented using a list of $J \times T$ integer variables, where variable $state(j \in J, t \in T)$ is defined as follows: if $x = 0$ then $state(j, t) = 0$ while if $x_{jtk} = 1$ then $state(j, t) = k$. Figure 1 shows an example of a solution and its graphical representation in the map. In this example, there are $J = 4$ facilities, $k = 3$ facility types, $T = 2$ periods of time, and $p_1 = 2$, $p_2 = 1$ and $p_3 = 1$. The demand nodes I changes over time. For each potential facility locations, the green and white squares indicate if a facility of type k can be located or not in such node.

For example, at node 2 and $t = 1$, any facility type can be opened. Node 3 at $t = 1$ is used with a facility of type 1 (types 1 and 3 were available). At $t = 2$, just type 3 is available for node 3, which is again used. Node 1 is never used. The plots below show the facilities and demand nodes in the map. The type of the opened facilities is depicted with circles with different radius (the coverage area).

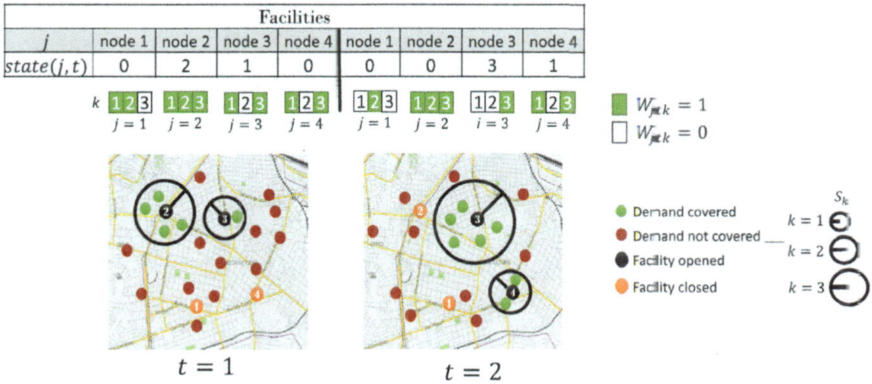

Fig. 1 Representation of a solution for DMCLP-FT

3 Two Illustrative Examples

In this section we show two examples of a problem with $I = 1500$, $J = 80$. The nodes are randomly taken from the TSP instance pm8079 [4]. We consider $K = 3$ types of facilities where the values of S_k depend on $maxD$ (defined as 50% of the maximal distance between all the facilities and demand nodes). Values are $S_1 = maxD \times 5\%$, $S_2 = maxD \times 10\%$, $S_3 = maxD \times 15\%$ The total number of facilities of each type to be located are $p_1 = 20$, $p_2 = 12$ and $p_3 = 8$. The demand values were randomly assigned using values in the interval [0, 100]. There are $T = 5$ periods of time. Each example has a different distribution of facility types in the periods.

As at this stage we are no looking for the best way to solve the problem, we apply a basic Hill climbing (HC) strategy [1] to explore how the solutions look like. The procedure for building the initial solution was based in the method described in [13], where the facilities to be opened are selected by roulette selection. We selected a mutation operator that randomly combines swap operator NSS1 and NSS2 [13] and roulette swap operator proposed in [6]. As an initial approach, we consider the dynamism as a sequence of several static problems. For each example we selected two of the best solutions obtained over 30 runs.

In the first example, the types available for each facility $j \in J$ are randomly assigned in every time period. For each period we consider 70% of J facilities of $k = 1$ type, 50% of $k = 2$ and 30% of $k = 3$. Figure 2 shows two solutions with a 97.46% (top) and 97.12% (bottom) of demand covered. For every period, the percentage of coverage and the facilities' locations are shown. The radius of every circle represents the facility type. The numbers highlighted in the examples represent the index (j) of the facilities.

Let's consider first the solution on top. As expected, a high level of coverage can be attained in different ways: using a few facilities with high coverage, or several

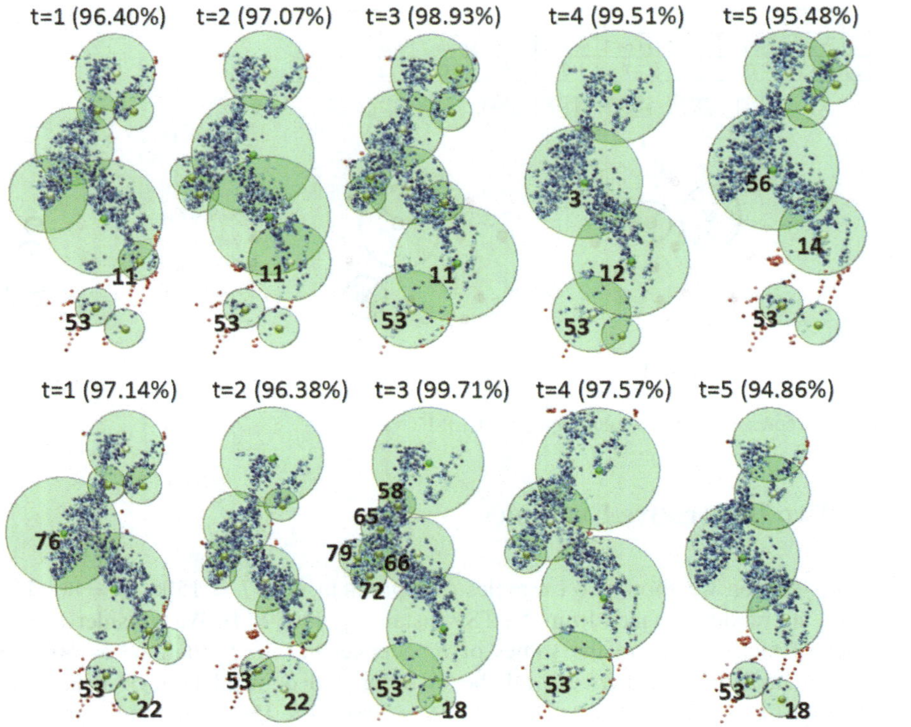

Fig. 2 Two solutions for the Example 1

ones with low or medium coverage ($k = 1, 2$) or a proper mix of them. Examples are in $t = 4$, $t = 1$ and $t = 2$ respectively. In $t = 4$, the facility $j = 3$ covers almost the same area of the facility $j = 56$ in the period $t = 5$. Both were opened with type $k = 3$. Facility $j = 11$ was used in $t = 1$, $t = 2$ and $t = 3$ but with different types $k = (1, 2, 3)$. The same area is covered by facility $j = 12$ in period $t = 4$ with $k = 3$, and by $j = 14$ in period $t = 5$ with $k = 2$. In this solution, the area around facilities $j = 53$ and $j = 14$ is hard to cover as there are few potential facilities locations in that zone and the demand nodes are quite sparsed.

The second solution (bottom) allows to observe other interesting features. The facility $j = 53$ is somehow complemented by $j = 22$ in periods $t = 1$ and $t = 2$, and by the facility $j = 18$ in periods $t = 3$ and $t = 5$, to cover the demand nodes located at the bottom of the map. We can also observe that facilities $j = 22$ and $j = 18$ are very close. The area covered by facility $j = 76$, $k = 3$ in period $t = 1$ is then covered by five facilities ($j = 58, 65, 66, 72, 79$) with $k = 1$ in period $t = 3$. These facilities are not open in others periods.

For the Example 2, we define a different distribution of the types available for each facility in every period. In $t = 1, 2$, all the facilities can be opened with any type; in $t = 3, 4$ the distribution of the available types is the same as in Example 1.

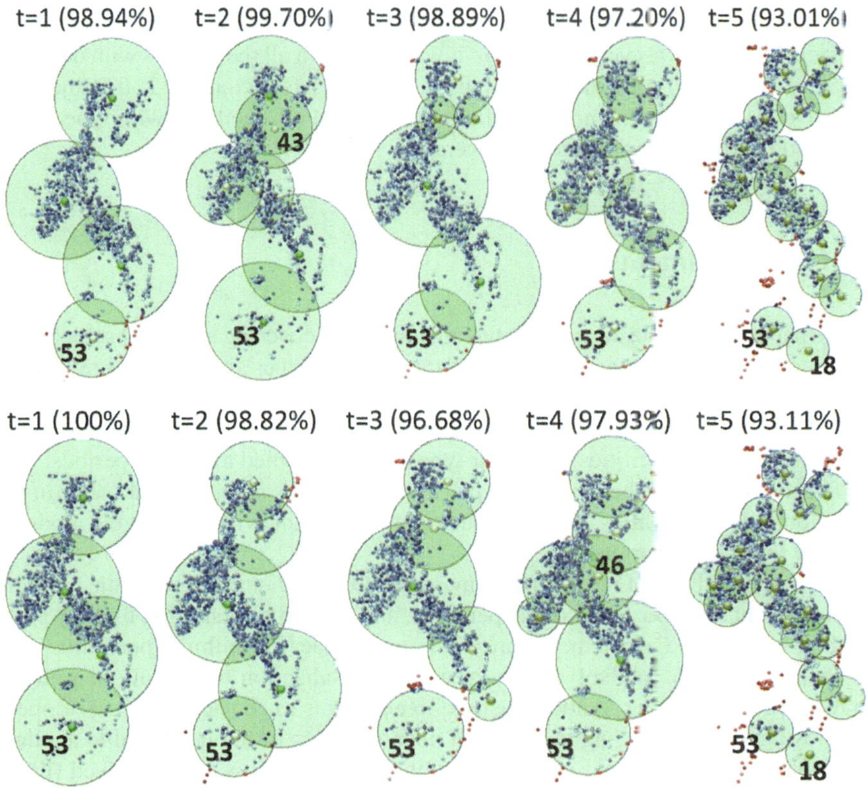

Fig. 3 Two solutions for the Example 2

Finally, at $t = 5$ the facilities can be opened just with $k = 1$. Figure 3 shows two solutions for the Example 2, with a 97.53% (top) and 97.29% (bottom) of covered demand.

For the top solution, it can be observed that the facility $j = 43$ in period $t = 2$ does not provide any additional coverage. The same situation happened in the bottom solution with the facility $j = 46$ in period $t = 4$, where such facility provided almost not additional coverage. In the bottom solution we note that the coverage achieved the 100% with four facilities with type $k = 3$. In both solution, at $t = 5$, the only two facilities available at the bottom ($j = 53$, 18) are opened.

We can see that the facilities with the type $k = 3$ were the most used in the first periods. The larger part of the facilities with type $k = 1$ was located in the period $t = 5$ to guarantee the covered demand. In the previous periods, the facility $j = 53$ was opened with the highest coverage radius and it could cover almost all demand nodes in that zone. At $t = 5$ both solutions just differ in four facilities. This period had the least covered demand, since the facilities only have available the type $k = 1$.

Some interesting features can be observed from these examples. In both of them the facility $j = 53$ was very important: it was opened in all the periods with different types: $(1, 1, 2, 2, 1)$ for Example 1 and $(3, 2, 2, 2, 1)$ for Example 2. It is the only facility that can cover the nodes in that zone. In general, facilities with type $k = 1$ appeared together to maximize the coverage.

4 Final Comments

In this paper a generalization of the dynamic maximal covering location problem was presented. Our model considers both different facility types and a dynamic variation of their availability over time. Using an illustrative example, we show the amount of information that can be gained analyzing the solution, not only in terms of coverage, but also in how the facilities (and their types) are distributed along the periods.

The distribution of the available types in the periods is an influential factor in the selection of open facilities. The previous examples are an evidence that the model is suitable for these situations.

Now, several lines of research are envisaged. One of them is to solve the problem in an efficient way. Based on our experience, and instead of searching for the "best" method to solve the problem, we are going to consider algorithmic portfolios. It is necessary to couple any solver chosen with a visualization tool to allow a proper analysis of the solutions. Another feature to explore is to consider the cost of opening/closing and/or changing the types of open facilities from one period to another. This cost may be neglected when the facility to locate can be easily moved (like a WiFi hotspots) it should be considered in other settings.

References

1. Bagherinejad, J., Seifbarghy, M., Shoeib, M.: Developing dynamic maximal covering location problem considering capacitated facilities and solving it using hill climbing and genetic algorithm. Eng. Rev. **37**(2), 178–193 (2017)
2. Church, R., ReVelle, C.: The maximal covering location problem. Pap. Reg. Sci. Assoc. **32**(1), 101–118 (1974)
3. Church, R.L., Murray, A.: Location Covering Models: History, Applications and Advancements. Advances in Spatial Science, 1ra edn. Springer International Publishing AG, New York (2018)
4. Cook, W.: The traveling salesman problem (TSP). Department of Combinatorics and Optimization, University of Waterloo (2019). http://www.math.uwaterloo.ca/tsp/index.html Accessed 14 July 2020
5. Heyns, A.M., van Vuuren, J.H.: Multi-type, multi-zone facility location. Geogr. Anal. **50**(1), 3–31 (2018)
6. Mišković, S.: A VNS-LP algorithm for the robust dynamic maximal covering location problem. OR Spectr. **39**(4), 1011–1033 (2017)
7. Moore, G.C., ReVelle, C.: The hierarchical service location problem. Manag. Sci. **28**(7), 775–780 (1982)

8. Murray, A.T.: Maximal coverage location problem: impacts, significance, and evolution. Int. Reg. Sci. Rev. **39**(1), 5–27 (2016)
9. Porras, C., Fajardo, J., Rosete, A.: Multi-coverage dynamic maximal covering location problem. Revista de Investigación Operacional (RIO) **40**(1), 140–149 (2019)
10. Reinelt, G.: TSPLIB. A traveling salesman problem library. ORSA J. Comput. **3**(4), 376–384 (1991)
11. Schilling, D.A.: Dynamic location modeling for public-sector facilities: a multicriteria approach. Decis. Sci. **11**(4), 714–724 (1980)
12. Stanimirović, Z., Mišković, S., Trifunović, D., Veljović, V.: A two-phase optimization method for solving the multi-type maximal covering location problem in emergency service networks. Inf. Technol. Control **46**(1), 100–117 (2017)
13. Zarandi, M.F., Davari, S., Sisakht, S.H.: The large-scale dynamic maximal covering location problem. Math. Comput. Model. **57**(3–4), 710–719 (2013)

Implicative Linguistic Summaries

M. Eugenia Cornejo and Jesús Medina

Abstract Linguistic summaries are a useful tool for extracting information of a database. In this paper, we show the usual linguistic summaries with their corresponding protoforms. In addition, we propose the notion of implicative linguistic summary and a method to compute its validity by using fuzzy implications.

Keywords Linguistic summary · Fuzzy set · Fuzzy implication

1 Introduction

Linguistic summaries arose as a friendly method for extracting the information contained in a database in a more understandable and comprehensible way [2–6]. Specifically, these summaries are linguistically quantified propositions, including helpful terms and ideas, in order to obtain an overview of a database and to reach the aims of the recipient of these summaries. In this paper, we review some concepts related to the usual types of linguistic summaries existing in the literature. Our goal in this work is to provide the notion of extended linguistic summary with an implicative character. Considering this approach, we introduce a new type of linguistic summary called implicative linguistic summary and we propose a formula to compute its validity.

Partially supported by the State Research Agency (AEI) and the European Regional Development Fund (ERDF) project TIN2016-76653-P.

M. E. Cornejo (✉) · J. Medina
Department of Mathematics, University of Cádiz, Cádiz, Spain
e-mail: mariaeugenia.cornejo@uca.es

J. Medina
e-mail: jesus.medina@uca.es

M. E. Cornejo et al. (eds.), *Computational Intelligence and Mathematics for Tackling Complex Problems 2*, Studies in Computational Intelligence 955,
https://doi.org/10.1007/978-3-030-88817-6_3

2 Linguistic Summaries

In this section, we will recall the approach proposed by Yager for summarizing data by means of linguistically quantified propositions [5]. This approach, which is based on fuzzy set theory, provides a useful tool for representing and manipulating the information contained in a database. First of all, we show the notion of linguistic summary.

Definition 1 Let D be a database composed of a finite set of objects B characterized by a finite set of attributes A. A linguistic summary of D is given by the tuple (S, Q, t), where:

(1) S is a summarizer, that is, a fuzzy predicate defined on the domain of a particular attribute in the database;
(2) Q is a fuzzy quantifier;
(3) t is the truth value or validity of the linguistic summary, that is, a value in the unit interval $[0, 1]$. The closer t to 1 is more truthful the proposed summary will be and the closer t is to 0 the less truthful it will be.

Linguistic summaries can also include a qualifier R, that is, another fuzzy predicate defined on the domain of an attribute (different from the considered attribute for the summarizer) determining a (fuzzy) subset of B.

Now, we present a simple context to illustrate the underlying idea of the notion of linguistic summary. For instance, it may be interesting for a manager to obtain the following summaries of a database containing information on the staff of his company: *"Most employees earn a high salary"*, *"Most employees are of medium age and earn a high salary"*, *"Most employees have a senior level of expertise or have a high qualification"*, *"Almost all young employees have a low qualification"*. Such linguistically quantified propositions, which can be identified with the notion of linguistic summary, may be of a great value for the company. For example, these linguistic summaries can be useful to establish some relation between seniority and salary. Therefore, it is important to know a procedure to compute their validity. Before showing a method to compute the validity of a linguistic summary, we will show the usual types of linguistic summaries existing in the literature: *simple* and *extended linguistic summaries*.

(1) **Simple linguistic summary:** An example of this type of linguistic summary is the following linguistically quantified proposition *"Most employees earn a high salary"*. Specifically, the summarizer S can be identified as the fuzzy predicate *"high salary"* and the fuzzy quantifier Q can be identified as the quantifier in natural language *"Most"*. Notice that, the summarizer S and the quantifier Q can be represented by using the notion of fuzzy set. On the one hand, the summarizer *"high salary"* is associated with the attribute $a =$ salary, whose possible set of values is V_a, and it is defined as the following fuzzy set:

$$S = \{(x, \mu_S(x)) \mid x \in V_a, \mu_S(x) \in [0, 1]\}$$

where μ_S is a membership function of S.

On the other hand, the fuzzy quantifier *"Most"* can be defined as the following fuzzy set:

$$Q = \{(r, \mu_Q(r)) \mid r \in X, \mu_Q(r) \in [0, 1]\}$$

where $X = [0, 1]$ and μ_Q is an increasing membership function of Q. Notice that, the domain of the fuzzy quantifier *"Most"* is the unit interval because it is a relative quantifier, that is, it indicates a proportion of elements with respect to the total. When we consider an absolute quantifier, which indicates a certain amount, then its domain is $\mathbb{R}^+$.

According to the notion of protoform (abstract prototype of a linguistically quantified proposition) of a linguistic summary proposed by Kacprzyk and Zadrozny [3], we can write a simple linguistic summary as follows:

$$Q \text{ objects in the database are } S$$

(2) **Extended linguistic summary:** The linguistically quantified proposition *"Almost all young employees have a low qualification"* is an example of extended linguistic summary. This fact is due to, as well as the summarizer S representing the fuzzy predicate *"low qualification"*, we can identify another fuzzy predicate defined on the domain of the attribute age determining a fuzzy subset of employees. Therefore, we are qualifying our objects which gives rise to denote this kind of fuzzy predicate as qualifier R. The qualifier *"young"* can be also defined as a fuzzy set associated with the attribute $a' =$ age, whose possible set of values is V'_a. Here, the fuzzy quantifier Q represents the quantifier in natural language *"Almost all"*. The protoform of an extended linguistic summary is given by the following expression:

$$Q \text{ of } R \text{ objects in the database are } S$$

Now, we will present a new notion of linguistic summary which arises from the notion of extended linguistic summary adding an implicative character. Originally, Yager [6] proposed that the protoform *"Almost all young employees have a low qualification"*, is equivalent to *"In almost all employees, to be young implies to have a low qualification"*. Indeed, we could write this protoform as *"In almost all employees, if the age is young then the qualification is low"* to express the same information that in the original proposition. However, the meaning of both expressions is different. The first summary express that, among all the young people, almost all of them have a low qualification, nevertheless the second or third protoform is focused on all employees and aims to explain that if one person is young then he/she has low qualification.

Therefore, this second point of view, capable of inferring a conclusion from a condition, needs to be interpreted by a new linguistic summary.

(3) **Implicative linguistic summary:** The reasoning above leads us to propose the next expression as the protoform of an implicative linguistic summary:

$$\textit{In } Q \textit{ objects in the database, if } R \textit{ then } S$$

Once we have presented the different types of linguistic summaries that we consider in this work, we will show how to compute the validity of each of them.

3 Computing the Validity of Linguistic Summaries

In this section, we present the procedure for computing the truth value or validity of the previous types of linguistic summaries. To reach this goal, we will consider Zadeh's fuzzy logic based calculus of linguistically quantified propositions [7] and the basic validity criterion introduced by Yager [5], which has been used by many authors later on due to its importance in the general framework.

Consider a finite database D whose set of objects $B = \{b_1, b_2, \ldots, b_n\}$ is described by a set of attributes $A = \{a_1, a_2, \ldots, a_m\}$. Suppose that each attribute a_j has associated a set of values V_{a_j} and a mapping $\bar{a}_j \colon B \to V_{a_j}$, which assigns the value $\bar{a}_j(b_i) = y_i \in V_{a_j}$ to each object b_i, for all $i \in \{1, \ldots, n\}$ and $j \in \{1, \ldots, m\}$. Considering this notation, we can express the database as follows:

$$D = \{[\bar{a}_1(b_i), \bar{a}_2(b_i), \ldots, \bar{a}_m(b_i)]\}_{i \in \{1,\ldots,n\}}$$

Given $j \in \{1, \ldots, m\}$, suppose that the summarizer S is a fuzzy predicate associated with the attribute a_j. Then, for each $i \in \{1, \ldots, n\}$, we compute the degree to which b_i satisfies the summarizer S, that is, $\mu_S(\bar{a}_j(b_i))$. In order to simplify the notation, we will write $\mu_S(y_i)$ instead of $\mu_S(\bar{a}_j(b_i))$. We will also consider this notation with respect to a possible qualifier in the linguistic summary. Following the idea proposed in [5, 6], we can obtain the validity of the previously mentioned linguistic summaries by means of the formulas given below, where $\wedge$ and $\vee$ can represent a t-norm and a t-conorm, respectively.

(1) **Validity of a simple linguistic summary:**

$$t(Q \textit{ objects in the database are } S) = \mu_Q\left(\frac{1}{n}\sum_{i=1}^{n} \mu_S(y_i)\right)$$

(2) **Validity of an extended linguistic summary:**

$$t(Q \text{ of } R \text{ objects in the database are } S) = \mu_Q\left(\frac{\sum_{i=1}^{r} \mu_R(y_i) \wedge \mu_S(y_i)}{\sum_{i=1}^{n} \mu_R(y_i)}\right)$$

In this paper, we will relate the implicative linguistic summary with the concept of IF-THEN rule. These rules allow us to represent knowledge in a quite appropriate way in the context of linguistics because they express the human empirical and heuristic knowledge underlying in our communication language. Since we interpret the implicative linguistic summary as an expression capable of inferring a conclusion from a condition, we propose to compute the validity of an implicative linguistic summary by using an implication operator.

(3) **Validity of an implicative linguistic summary:**

$$t(\text{In } Q \text{ objects in the database, if } R \text{ then } S) = \mu_Q\left(\frac{1}{n}\sum_{i=1}^{n} \mu_R(y_i) \rightarrow \mu_S(y_i)\right)$$

Following the philosophy of classical logic, since we are interpreting the implicative linguistic summary as an IF-THEN rule, it seems natural to require that $\mu_R(y_i) \rightarrow \mu_S(y_i) = 1$ when the antecedent $\mu_R(y_i)$ is false, that is, $\mu_R(y_i) = 0$. Therefore, all objects will be considered in the computation of the validity of the implicative linguistic summary, including the objects that do not satisfy the antecedent, which have also an important role. This fact also justifies the division by the number of objects in the database, instead of dividing by the number of objects satisfying R. In addition, it is also important to require monotonicity properties to the implication operator. Specifically, $\rightarrow$ should be decreasing in the antecedent and increasing in the consequent. Therefore, we propose that the operator $\rightarrow$ considered in the formula for computing the validity of an implicative linguistic summary be a fuzzy implication.

Finally, we will include an example in order to illustrate the proposed approach to compute the validity of an implicative linguistic summary.

Example 1 Consider a database D containing the information on the staff of a certain company. This company has 100 employees and the information in which it is interested in is given in Table 1. Taking into account that the age up to 35 years is considered *"young"* and the salary up to 1100 euros is considered *"low salary"*, from the information contained in Table 1, we can establish the following linguistic summaries: *"Most young employees earn a low salary"*, and *"In most cases, if the age is young then salary is low"*. Clearly, the first expression is identified with an extended linguistic summary and the second one with an implicative linguistic summary. Notice that, the summarizer S represents the fuzzy predicate *"low*

Table 1 Information on the employees

N° employees	Age	Salary	Seniority	Qualification
10	[18, 32]	[500, 1100]	Low	Low
4	[28, 35]	[1200, 1500]	Normal	High
53	[40, 55]	[1500, 1800]	High	High
33	[50, 65]	[1800, 2500]	High	Normal

$N^\circ Employees$	Age	μ_R	$Salary$	μ_S
2	23	0.92	865	0.659
1	25	0.778	900	0.5
2	28	0.435	1050	0.031
1	30	0.222	955	0.263
1	30	0.222	1015	0.09
1	31	0.142	970	0.211
2	32	0.08	900	0.5
2	32	0.08	1200	0
2	32	0.08	1350	0
86	[40, 65]	0	-	-

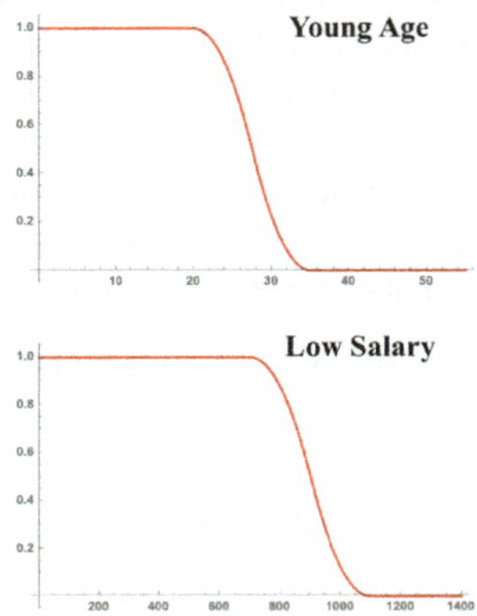

Fig. 1 Information on young employees (left side) and μ_R, μ_S mappings (right side)

salary" and the qualifier R represents the fuzzy predicate *"young"*. The membership functions associated with S and R are depicted in Fig. 1. The fuzzy quantifier Q represents the quantifier *"Most"*. The membership function associated with Q is $\mu_Q\colon [0,1] \to [0,1]$ defined as $Q(r) = r^2$. In order to compute the validity of these two linguistic summaries, we consider the Łukasiewicz t-norm $\wedge_{\text{L}}$ and the Łukasiewicz implication $\to_{\text{L}}$, this last one as the fuzzy implication of the implicative summary.

The table included in Fig. 1 does not contain the information corresponding to employees whose age range is between 40 and 65 years since, regardless of the salary earned by these employees, $\mu_R(y_i) \to_{\text{L}} \mu_S(y_i) = 1$. This fact is due to the membership function μ_R takes the value 0 for these employees.

Now, we can emphasize the motivation to require that the implication operator be decreasing in the antecedent and increasing in the consequent. For instance, there are an employee with 25 years old and another employee with 30 years old whose salaries coincide, that is, they earn 900 euros. Applying the membership functions, we have that $\mu_R(32) \leq \mu_R(25)$ and therefore, we obtain that $0.722 = \mu_R(25) \to_{\text{L}} \mu_S(900) \leq \mu_R(32) \to_{\text{L}} \mu_S(900) = 1.0$. This result makes sense since, according to the statement *"if the age is young then salary is low"*, one expects that an employee with 25 years old earn a lower salary than another employee with 30 years old.

On the other hand, from the table included in Fig. 1, we can see that there are two employees with 30 years old earning different salaries. Following a similar reasoning to the one given previously, one expects that the value increases when the salary decreases. This fact is supported by the following inequality $0.868 = \mu_R(30) \rightarrow_L \mu_S(1015) \leq \mu_R(30) \rightarrow_L \mu_S(955) = 1.0$, where $\mu_S(1015) \leq \mu_S(955)$.

Finally, from the information contained in Fig. 1, we obtain:

$$t(Q \text{ of } R \text{ employees are } S) = \mu_Q\left(\frac{\sum_{i=1}^{100} \mu_R(y_i) \wedge_L \mu_S(y_i)}{\sum_{i=1}^{100} \mu_R(y_i)}\right) = 0.315$$

$$t(\text{In } Q \text{ employees, if } R \text{ then } S) = \mu_Q\left(\frac{1}{100}\sum_{i=1}^{100} \mu_R(y_i) \rightarrow_L \mu_S(y_i)\right) = 0.97$$

Therefore, the validity of the linguistic summary *"Most young employees earn a low salary"* is lesser than the truth value of *"In most cases, if the age is young then salary is low"*. This result shows the suitability of our proposal, since of the 14 young employees, we have that 10 earn a low salary, so the truth value associated with this information should be high. □

4 Conclusions and Future Work

The main contribution of the paper has consisted in introducing the notion of implicative linguistic summary which is needed to inferring a conclusion from a condition. Moreover, we have shown an original formula to compute the validity of implicative linguistic summaries. As a future work, we will study alternative formulas for the computation of the validity of an implicative linguistic summary, for example, considering residuated implications [1] and/or taking into account different normalization factors.

References

1. Cornejo, M.E., Medina, J., Ramírez-Poussa, E.: Multi-adjoint algebras versus non-commutative residuated structures. Int. J. Approx. Reason. **66**, 119–138 (2015)
2. Kacprzyk, J., Yager, R.R.: Linguistic summaries of data using fuzzy logic. Int. J. Gen. Syst. **30**(2), 133–154 (2001)
3. Kacprzyk, J., Zadrozny, S.: Linguistic database summaries and their protoforms: towards natural language based knowledge discovery tools. Inf. Sci. **173**(4), 281–304 (2005). Dealing with uncertainty in data mining and information extraction

4. Kacprzyk, J., Yager, R.R., Zadrozny, S.: A fuzzy logic based approach to linguistic summaries in databases. Int. J. Appl. Math. Comput. Sci. **10**, 813–834 (2000)
5. Yager, R.R.: A new approach to the summarization of data. Inf. Sci. **28**(1), 69–86 (1982)
6. Yager, R.R., Yager, R.L.: Using linguistic summaries and concepts for understanding large data. Eng. Appl. Artif. Intell. **56**(C), 273–280 (2016)
7. Zadeh, L.A.: A computational approach to fuzzy quantifiers in natural languages. Comput. Math. Appl. **9**(1), 149–184 (1983)

Comparison of Discrete Memetic Evolutionary Metaheuristics for TSP

Gergö Fogarasi, Boldizsár Tüü-Szabó, Péter Földesi, and László T. Kóczy

Abstract In our paper we compare discrete memetic evolutionary metaheuristics (and other algorithms) which are applicable (also) for the widely studied and industrially applied (symmetric, Euclidean) NP-hard combinatorial optimization problem called Traveling Salesman Problem (TSP) such as DBMEA (Discrete Bacterial Memetic Evolutionary Algorithm), DMTLBO (Discrete Memetic Teaching–Learning Based Optimization) not to mention DMSSA (Discrete Memetic Squirrel Search Algorithm) algorithms. The comparisons occurred under the same fixed conditions.

Keywords TSP · DBMEA · DMTLBO · DMSSA

1 Introduction

The Traveling Salesman Problem (TSP) was first studied by Karl Menger in 1930s, and nowadays is one of the most widely studied combinatorial optimization problems. The general form of TSP involves a salesman who starts the journey from the company's headquarters, visits each city at least (and preferable at most) once, and then he returns to the starting place. The task is to find the route with minimal cost (minimum time spent or distance travelled) [8].

G. Fogarasi · B. Tüü-Szabó (✉) · P. Földesi · L. T. Kóczy
Szechenyi Istvan University, GyOr 9026, Hungary
e-mail: tuu.szabo.boldizsar@sze.hu

G. Fogarasi
e-mail: fogarasi.gergo@sze.hu

P. Földesi
e-mail: foldesi@sze.hu

L. T. Kóczy
e-mail: koczy@sze.hu

L. T. Kóczy
Budapest University of Technology and Economics, Budapest 1117, Hungary

M. E. Cornejo et al. (eds.), *Computational Intelligence and Mathematics for Tackling Complex Problems 2*, Studies in Computational Intelligence 955,
https://doi.org/10.1007/978-3-030-88817-6_4

This problem has several application areas, such as logistics, vehicle routing, planning, and the manufacture of microchips [8].

The Traveling Salesman Problem can be defined as a graph search problem with edge weights. In this case the vertices represent the cities while the edges represent the roads between the cities and the edge weights represent the lengths of the roads (or the appropriate amount of time or cost to take the given road) [8].

In case of the original version of TSP, the graph is complete, Euclidean, undirected, symmetric. The goal is to find the shortest Hamiltonian cycle, or as an alterna-tive wording, find an optimal permutation of the vertices with a minimal total cost [8].

Due to the fact that we do not know a polinomial solution exists, currently existing exact algorithms' worst-case time complexity is exponential. Consequently, if we want quick results for large instances, only approximate solutions can be the solution (for example memetic metaheuristics)—fortunately, their results are usually very close to the optimum values, and in most of the applications it is enough [8].

In our previous research we compared discrete memetic evolutionary metaheuristics which are applied for TSP, but with fewer algorithms [3].

2 Increasing Number of Existing Metaheuristics

A lot of new (nature inspired) metaheuristics are published year after year, some of them have been applied to discrete problems (including TSP), while the rest has not been investigated for TSP (but this could be possible) [1]. Often these algorithms are the accusation that there is no mathematical novelty in them, they only illustrate an already known algorithm with an alternative natural phenomenon—however, in many cases there is a real mathematical innovation in them (Fig. 1).

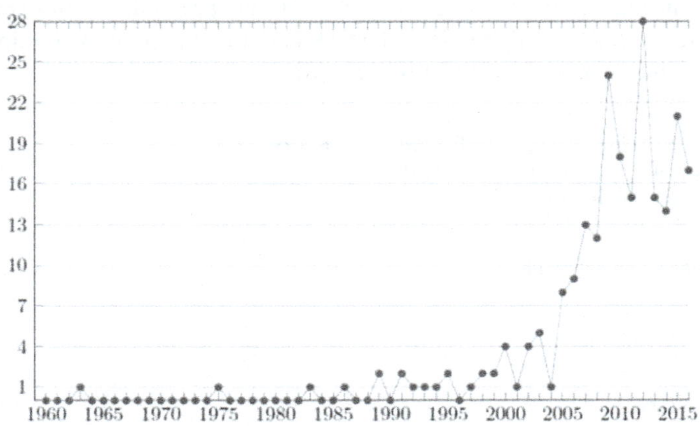

Fig. 1 Increasing number of newly published metaheuristics [1]

3 Common Parts of the Algorithms

There are many common elements in the algorithms, and if they were programmed in different ways, we would get misleading results, since these are not the things that should decide the difference between the algorithms (for example processing of the instance file, runtime measurement, local search commonly used by many algorithms).

Fortunately, in case of TSP, it is possible to generate initial solutions effectively (in a relatively short time) which fulfills the conditions but not necessarily optimal (unlike in case of some specific versions of the TSP)—they can be used as initial individuals of a population-based algorithm.

This is done by applying different heuristics, such as Nearest Neighbor (NN), Second Nearest Neighbor (2NN or SNN), Alternating Nearest Neighbor (ANN), Random Alternating Nearest Neighbor (RANN) or Random Heuristic (RH).

There are many TSP operations which are commonly used by many metaheuristics. These include the global and local search techniques (2-opt and 3-opt were used).

4 Analized Discrete Memetic Evolutionary Metaheuristics and Other Algorithms, Which Are Applied for TSP

In this research, 12 algorithms have been programmed, then the measurement results were made.

DBMEA (Discrete Bacterial Memetic Evolutionary Algorithm) is a population based memetic evolutionary algorithm. Unlike the non-discrete version, it is not using gradient type local search but 2-opt and 3-opt local searches [8, 9] (Fig. 2).

DMTLBO (Discrete Memetic Teaching–Learning Based Optimization) is a discretized and memetic version of the previously existing Teaching–Learning Based Optimization and it is described in our previous work [3].

DMSSA (Discrete Memetic Squirrel Search Algorithm) is the discretized, memetic version [3] of a nature-inspired optimization paradigm, named as squirrel search algorithm (SSA). This optimizer imitates the dynamic foraging behaviour of flying squirrels and their efficient way of locomotion known as gliding. Gliding is an effective mechanism for travelling long distances. SSA mathematically models this behaviour to realize the process of optimization. [6] DMSSA is able to use for TSP [3].

DMPPOA (Discrete Memetic Predator–Prey Optimization Algorithm) is a relatively new metaheuristic for TSP, first described in our previous work [3].

ACO (Ant Colony Optimization) [10] is a metaheuristic which is inspired by the behaviour of real ants to forage for food. This metaheuristic has been widely used in many optimization problems and fields [2].

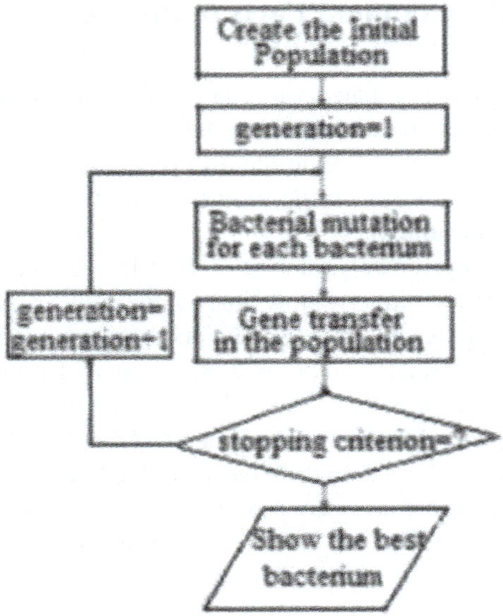

Fig. 2 Flow chart of DBMEA [8]

Concorde, the Original Lin-Kernighan Heuristic and LKH 3.0.6 (an effective implementation of the Lin-Kernighan TSP Heuristic by Keld Helsgaun) [4, 5] are three really important algorithms, they are all inevitable on this subject.

2-opt Approximation, Greedy with Farthest Insertion, Iterated Hill Climbing, and SA (Simulated Annealing) [10] are very outdated, obsolete techniques, yet they are all included in the comparison for a fuller picture, but good results are not expected.

SA is not based on any of the classical optimization techniques but is based on heuristics from annealing process. This heuristic method is based on the motivation from how crystalline structures are formed from annealing process. SA is similar to Hill Climbing or Gradient Search with a few modifications [15].

5 The Applied Simulation Test Methods

All of the analyzed algorithms are written in C++. To compare the simulation results and runtimes, the same PC and operating system have been used. The runtime-measurement was integrated into the software. The measurements were done several times to try to avoid, as far as possible, to draw conclusions from rare extreme cases.

6 TSP Instances

There is a necessity for test instances when comparing TSP algorithms. To do this, we used instances written in the most common file format (TSPLIB format), from the 5 most widely used TSP instance collections in the area. The total number of instanc-es on these websites is 243, of which only a few have been used [11–14, 18]. In case of the TSPLIB format, the edge weights are given only implicitly (since each vertex is located on the same two-dimensional plane, the edge weights are calculated from the two-dimensional coordinates of the graph vertices).

7 Results of the Simulation and Its Conclusions

Most important elements of the simulation study were the discrete memetic meta-heuristics. It can be said generally that from these, DBMEA is the best one. However, ACO [10] is really close to it, besides sometimes even better. For example, in terms of pma343 instance, ACO is producing a better result than DBMEA, moreover it is faster. This could be just a unique case, maybe not. Further research is needed.

The well-known algorithm named Concorde always gives the optimal tour length, however it is really slow compared to metaheuristics. Due to its slowness in case of large instances, it cannot be applied effectively, other methods are needed (for example DBMEA, ACO or LKH improved by Keld Helsgaun)

LKH 3.0.6 gives the second best results after Concorde, and it is much faster.

To reach a clear overall picture, we also used classic gradient-based and other types of obsolete algorithms. They have produced bad results, as expected (Tables 1 and 2).

8 Future Research Opportunities

The ACO [10] algorithm needs to be examined more profoundly. For example, in terms of pma343 instance, ACO is producing a better result than DBMEA, moreover it is faster. This could be just a unique case, maybe not. Further research is needed.

For a more comprehensive comparison, more algorithms (memetic metaheuristics, and also other types) have to be examined, with various versions of parameters (this could cause changes in results and in runtimes). For instance: Branch and Bound, Ant Lion and Firefly Algorithm (two Insect Based Algorithms) [16], CSO [7], BCO, Artificial Bee Colony, FUNGI [17], Tabu Search, other versions of Simulated Annealing [10, 5], PSO [19], Greedy Randomized Adaptive Search Procedure, Iterated Local Sarch, CSO, DSOA [20], Bat-Inspired Algorithm, Grey Wolf Optimizer.

A new possible way to continue the research could be to explore the relationship between different graph theory parameters of the instances and the behaviour of the

Table 1 First half of the results

TSP instance	source	number of nodes	DBMEA(Discrete Bacterial Memetic Evolutionary Algorithm)		DMTLBO (Discrete Memetic Teaching-Learning Based Optimization)		DMSSA (Discrete Memetic Squirrel Search Algorithm)		DMPPOA (Discrete Memetic Predator-Pray Optimization Algorithm)		ACO (Ant Colony Optimization)		Concorde (Dec 19.. 2003)	
			value	runtime [s]	value	runtime [s]	value	runtime [s]	value	runtime [s]	value	runtime [s]	value	runtime [s]
wi29.tsp	2	29	N/A	N/A	27601.2	6.653441	28145.2	0.0719972	27601.2	42.916	27603	1.97	27603	0.08
dj38.tsp	2	38	N/A	N/A	6659.43	8.588706	6659.43	0.0487136	6659.43	44.41	6656	0.85	6656	0.06
att48.tsp	1	48	N/A	N/A	33523.7	10.861869	40526.4	0.0222358	33523.7	160.552	10628	2.67	10628	0.24
eil51.tsp	1	51	428.872	0.395461	428.982	11.813101	438.605	0.123465	428.982	135.984	426	3.01	426	0.06
st70.tsp	1	70	677.11	0.170833	677.11	15.415072	751.273	0.0919927	677.11	127.519	675	2.6443	675	0.14
ch130.tsp	1	130	6110.72	1.37152	6163.49	27.694073	6673.69	0.819529	6135.67	211.1	6110	6.75	6110	0.46
pr144.tsp	1	144	58535.2	> 3600	58535.2	26.0698427	61650.7	0.214415	58535.2	150.514	58537	4.32	58537	0.7
kroA150.tsp	1	150	26524.9	2.299412	26696.5	32.169267	27961.9	1.261174	26798.2	1611.7102	26524	7.83	26524	1.76
kroB150.tsp	1	150	26127.7	2.513768	26227.1	31.400193	27705	1.30867	26158.8	140.99	26130	5.61	26130	2.64
pma343.tsp	4	343	1387.51	20.0484799	1399.2	64.275676	1490.2	9.530591	1393.4	1490.741336	1368	2.17	1368	7.7
rbu737.tsp	4	737	3354.90	212.469464	3391.49	247.833173	3640.15	165.352175	3633.64	> 3600	3358	4.032	N/A	N/A
vm1084.tsp	1	1084	239570	794.690294	N/A	N/A	N/A	N/A	261865	> 3600	242559	87.15	239297	296.52
xpr2308.tsp	4	2308	7340.33	8325.0876421	N/A	N/A	8048.54	4334.350313	N/A	N/A	7221	503.27	N/A	N/A

Table 2 Second half of the results

TSP instance	source	number of nodes	LKH3.0.6. by Keld Helsgaun		original Lin-Kernighan (part of Concorde software package)		2-opt Approximation		Greedy with farthest insertion		Iterated Hill Climbing		SA (Simulated Annealing)	
			value	runtime [s]	value	runtime [s]	value	runtime [s]	value	runtime [s]	value	runtime [s]	value	runtime [s]
wi29.tsp	2	29	27603	0	27603	0.01	73551.8	0.007	82135	0.03	67895.3	2.57	29032	0.11
dj38.tsp	2	38	6656	0.01	6656	0.01	30517.5	0.004	8093	0.12	28330.3	2.96	8017	0.01
att48.tsp	1	48	10628	0	10628	0.04	10834.6	0.005	12431	0.025	10373.9	3.16	11003	0.02
eil51.tsp	1	51	426	0.01	426	0.02	515.943	0.005	511	0.67	329.641	0.8	511	3.17
st70.tsp	1	70	675	0.01	675	0.03	55562	0.005	961	0.01	862.02	3.2	985	2.11
ch130.tsp	1	130	6110	0.06	6110	0.07	8454.43	0.006	7013	0.12	6433.62	4.1	8059	0.18
pr144.tsp	1	144	58537	0.51	58537	0.1	158569	0.007	72131	0.039	17265.2	3.157	62511	2.99
kroA150.tsp	1	150	26524	0.1	26524	0.08	74400.23	0.006	32511	0.79	32115	15.083	43638	4.01
kroB150.tsp	1	150	26132	0.07	26132	0.09	72410.59	0.006	28011	0.85	28115	12.301	28059	5.12
pma343.tsp	4	343	1368	3.5	1369	0.29	64693	0.006	1679.24	0.258	1679.24	2.97	1679.24	3.17
rbu737.tsp	4	737	N/A	N/A	3316	0.31	69965.26	0.032	67827	0.018	64450	3.01	67827	4.03
vm1084.tsp	1	1084	239297	61.99	239462	0.89	349918	0.057	281120	0.015	279147	41.244	N/A	N/A
xpr2308.tsp	4	2308	7219	99.02	7245	1.11	89007	0.007	114531	03.jan	N/A	N/A	83042	2.97

algorithms; or investigate around a new idea of a novel type of local search technique, which could be integrated into the metaheuristics to replace (or supplement) 2-opt and 3-opt; or to use alternative methods of comparison of the results.

Acknowledgements This research was sponsored by: "EFOP-3.6.1-16-2016-00017 Internationalization, initiatives to establish a new source of researchers and graduates, and development of knowledge and technological transfer as instruments of intelligent specializations at Szechenyi University." and the number K124055 OTKA project.

References

1. Diaz, J.G.: Busqueda basada en estructura para problemas NP-Dificiles (Doctorado en Ciencias de la ComputaciOn). Instituto Politecnico Nacional, Centro de InvestigaciOn en ComputaciOn, Mexico (2017)
2. Donis-Diaz, C., Bello, R., Kacprzyk, J.: Using Ant Colony Optimization and Genetic Algorithms for the Linguistic Summarization of Creep Data (2015). https://doi.org/10.1007/978-3-319-11313-5_8
3. Fogarasi, G., Tüü-Szabó, B., Foldesi, P., Koczy, L.T.: Az Utazo Ugynok Problemara alkalmazhato diszkret memetikus evolucios metaheurisztikak osszehasonlftasa. LIM. https://doi.org/10.29177/LIM.2019.1.15 (2019).
4. Helsgaun, K.: An effective implementation of the Lin-Kernighan traveling salesman heuristic. Eur. J. Oper. Res. **126**, 106–130 (2000). https://doi.org/10.1016/S0377-2217(99)00284-2
5. Helsgaun, K.: An effective implementation of K-opt moves for the Lin- Kernighan TSP heuristic (2018)
6. Jain, M., Singh, V., Rani, A.: A novel nature-inspired algorithm for optimization: squirrel search algorithm. Swarm Evol. Comput. (2018). https://doi.org/10.1016/j.swevo.2018.02.013
7. Jati, G.K., Manurung, H.M., Suyanto: Discrete cuckoo search for traveling salesman problem. In: 7th International Conference on Computing and Convergence Technology (ICCCT), pp. 993–997 (2012)
8. Koczy, L.T., Foldesi, P., Tüü-Szabó, B.: A discrete bacterial memetic evolutionary algorithm for the traveling salesman problem, pp. 3261–3267 (2016). https://doi.org/10.1109/CEC.2016.7744202
9. Koczy , K.T., Foldesi, P., Tüü-Szabó, B.: Enhanced discrete bacterial memetic evolutionary algorithm - an efficacious metaheuristic for the Traveling sales-man optimization. Inf. Sci. (2017). https://doi.org/10.1016/j.ins.2017.09.069
10. Mica, O.: Comparison of metaheuristic methods by solving travelling salesman problem. In: Proceedings of the 9th International Scientific Conference INPROFORUM: Common Challenges - Different solutions - Mutual dialogue. Ceske Budejovice: Jihoceska uni- verzita v Ceskych Budejovidch, pp. 116–120. ISBN 978–80–7394–536–7 (2015)
11. Mona Lisa TSP Challenge, http://www.math.uwaterloo.ca/tsp/data/ml/monalisa.html, last accessed 2019/02/01+
12. National Traveling Salesman Problems, http://www.math.uwaterloo.ca/tsp/world/countries.html, last accessed 2019/02/01
13. Romanian and Swiss TSP instance, http://cadredidactice.ub.ro/ceraselacrisan/cercetare/, last accessed 2019/02/01
14. Ruprecht-Karls-Universitat Heidelberg, TSP instances, https://www.iwr.uni-heidelberg.de/groups/comopt/software/TSPLIB95/tsp/, last accessed 2019/02/01
15. Seshadri, A.: Simulated annealing for travelling salesman problem (2006)
16. Sharma, M., Sharma, S., Singh, G., Singh, R.: A study of insect based optimization algorithms: ant lion optimizer and firefly algorithm. Int. J. Eng. Sci. (IJOES) **26** (2017)

17. Tormasi, A., Koczy, L.T.: Experimenting with a new population-based optimization technique: FUNgal growth inspired (FUNGI) optimizer. Studies in Fuzziness and Soft Computing, pp. 123–135 (2018). https://doi.org/10.1007/978-3-319-75408-6_11
18. VLSI Data Sets, http://www.math.uwaterloo.ca/tsp/vlsi/index.html, last accessed 2019/02/01
19. Yan, X., Zhang, C., Luo, W., Li, W., Chen, W., Liu, H.: Solve traveling salesman problem using particle swarm optimization algorithm. Int. J. Comput. Sci. Issues (IJCSI), pp. 264–27 (2012)
20. Yong, W., Tao, W., Cheng-Zhi, Z., Hua-Juan, H.: A new stochastic optimization approach: dolphin swarm optimization algorithm. Int. J. Comput. Intell. Appl. 15, 1650011 (2016). https://doi.org/10.1142/S1469026816500115

Syntactic Analysis of Sentences Using Deep Neural Networks

David Muñoz-Valero, Luis Rodriguez-Benitez, Luis Jimenez-Linares, and Juan Moreno-Garcia

Abstract Deep learning is a very common tool for natural language processing. In this document the use of neural networks is proposed to perform the syntactic analysis of sentences. We have experimented with three different models of artificial neural networks to identify the subject and the predicate of a sentence. A test has also been conducted to determine the type of words that make up the sentence. The obtained results are different for each network tested: the first one obtains inadequate results, the second one reaches 60% success in the detection of the subject and the predicate, and the third achieves 100% in the detection of the subject and the predicate and identifies the type of word in 92% of cases.

Keywords Natural language processing · Deep learning · LSTM · seq2seq

1 Introduction

Artificial Intelligence has experienced a great increase in popularity in recent years due to the improvement in computing capacity and the large amount of information available. It has been applied in a wide variety of different research areas, such as

---

Supported by Information and System Technologies Department and Project TIN2015-64776-C3-3-R of the Spanish Ministry of Science and Innovation, co-funded by the European Regional Development Fund.

D. Muñoz-Valero · J. Moreno-Garcia (✉)
Universidad de Castilla-La Mancha, Escuela de Ingeniería Industrial, Toledo, Spain
e-mail: juan.moreno@uclm.es

D. Muñoz-Valero
e-mail: david.munoz6@alu.uclm.es

L. Rodriguez-Benitez · L. Jimenez-Linares
Universidad de Castilla-La Mancha, Escuela Superior de Informática, Ciudad Real, Spain
e-mail: luis.rodriguez@uclm.es

L. Jimenez-Linares
e-mail: luis.jimenez@uclm.es

M. E. Cornejo et al. (eds.), *Computational Intelligence and Mathematics for Tackling Complex Problems 2*, Studies in Computational Intelligence 955,
https://doi.org/10.1007/978-3-030-88817-6_5

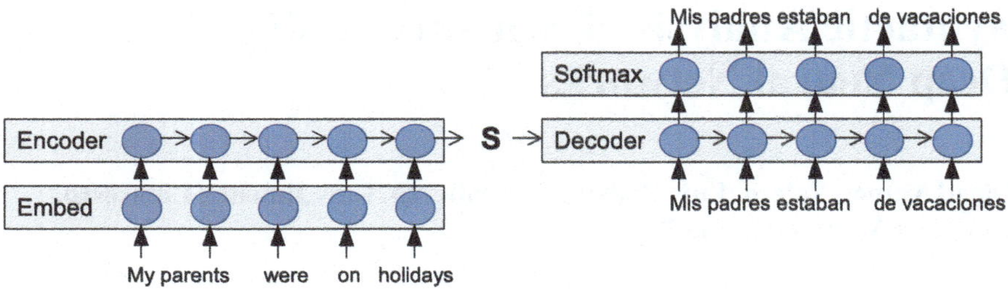

Fig. 1 seq2seq architecture

finance [1], medicine [2], virtual assistants such as Google Assistant, music [3], etc. Natural language processing (NLP) is one of the areas of research where it is most used today. Specifically, it is intended that a machine be able to understand the natural language from different sources (image, sound or text) [4]. Two relevant outstanding research fields within the NLP are *sentiment analysis* and *machine translation*. The objective of this work is to apply deep neural networks to the syntactic analysis of sentences. We present a study about the possibilities that some models of deep neural networks offer to perform a syntactic analysis, combining syntax and grammar [5]. Syntax is the part of grammar that studies the way in which words and groups are combined to express meanings along with the relationships that are established between all those units. To develop the neural network models, Tensorflow together with the use of the Keras high-level API are used. The rest of this paper is organised as follows: Sect. 2 presents a description of the proposed models to carry out the syntactic analysis. Section 3 details the obtained results from the tests. Finally, in Sect. 4, some conclusions and future works are exposed.

2 Detection of Sentence Components Using Deep Learning Models

Three models have been tested sequentially. This means that the use of a model was first decided, it was tested, and based on the results obtained and new research, the next one to be tested was decided. The models tested have been *seq2seq*,[1] a model based on the one proposed by George-Bogdan Ivanov[2] to perform "Part-Of-Speech tagging", and finally a combination of both has been designed.

The first model is called *seq2seq* and is also known as Encoder-Decoder [6]. It has been selected because it is used in *machine translation* obtaining very good results. It has as major advantage that it can have a different size for the input and for the output sequences, also allowing the logical order of the input elements (words in our case) to be varied in both sequences. Figure 1 shows its architecture. As it can be observed,

[1] https://keras.io/examples/lstm_seq2seq/.

[2] https://nlpforhackers.io/lstm-pos-tagger-keras/.

Table 1 Tags used in this document

Tag	Meaning	Tag	Meaning
CC	Coordinating conjunction	POS	Possessive ending
CD	Cardinal number	PRP	Personal pronoun
DT	Determiner	PRP$	Possessive pronoun
IN	Preposition or subordinating conjunction	TO	to
JJ	Adjective	VB	Verb base form
NN	Noun singular or mass	VBD	Verb past tense
NNS	Noun plural	VBN	Verb past participle
NNP	Proper noun singular	VBP	Verb non-3rd person singular present
NNPS	Proper noun plural		

it has an input sequence that is received by the encoder, it sequentially transfers it to the decoder that generates the translation as output in the target language, in this case Spanish.

The second model consists of several layers and it is called $1-LSTM$. One of the most relevant is a bi-directional long-short memory (LSTM) that allows the network to remember characteristics of the data that arrived before the word that is being processed at that moment. The bidirectionality of this layer allows it to take into account the characteristics of the word that arrived in past and subsequent iterations. The modification has consisted in preparing the network to recognise subjects and predicates. The network training takes as input the labels that are intended for each word of the sentence. Once the network has been trained it takes as input the words of the sentence to be evaluated. For example, the labels "SS" and "PR" are used to indicate that a word is part of the subject or of the predicate respectively. In the sentence "The dog chased the cat" the input and output sequences are ['The', 'dog', 'chased', 'the', 'cat'] and ['SS', 'SS' , 'PR', 'PR', 'PR'] respectively.

In view of the obtained results with the previous model, it was decided to combine these two networks to study its results with the aim of constructing a syntactic tree that indicates the belonging to the subject or the predicate of each word as well as its type. Table 1 shows the meaning of each tag assigned to the words in the examples in this document. The first network performs the task of "part-of-speech tagging" and the second identifies the subjects and predicates of the sentence. So, the second network uses the output of the first as input data. Figure 2 shows the scheme of this model called $2-LSTM$. Both networks are trained independently. As an example here-under details the training of the sentence "The dog chased the cat":

First network: It encodes the input sequence as ['The', 'dog', 'chased', 'the', 'cat'], obtaining as output ['DT', 'NN', 'VBD', 'DT', 'NN'].

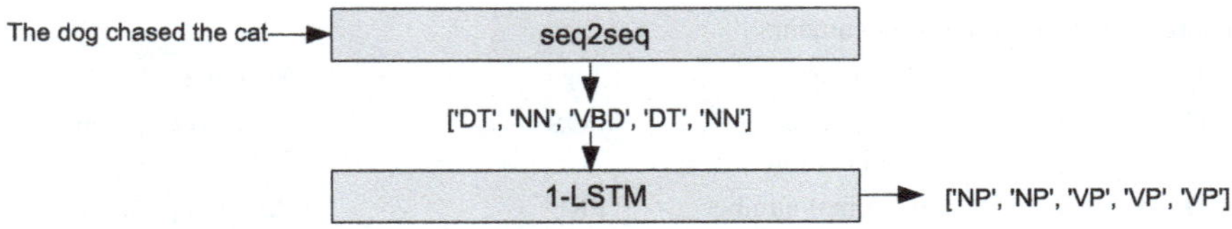

Fig. 2 $2 - LSTM$ architecture

Second network: It takes the output of the previous network generating a sequence of output labels, in this case ['NP', 'NP', 'VP', 'VP', 'VP'].

$2 - LSTM$ can be used when it has been trained. The sentence must be used as an input to the first network. As you can see, this model detects if a word belongs to the subject or the predicate, and also classifies the type of word it is. Next section shows the results obtained using each of these networks.

3 Experimental Results

seq2seq network has been trained using a set consisting of 2500 samples. A first test is carried out to verify the operation of the networka and it consists in the analysis of some sentences taken from NLTK library Penn Treebank Corpus.[3] Table 2 shows the results obtained for three sentences of the studied ones. The first sentence analysed has as subject *Pierre Vinken 61 years old* that does not correspond to the one detected by the network (Table 2). The second sentence has not been classified correctly, generating again a very long subject. Finally, the third sentence has a behaviour similar to the previous ones, the subject is very long. As a conclusion it can be said that the model is not able to correctly identify the subjects and predicates of the sentences that are supplied to it. It is a major problem considering that these examples belong to the training set. As a positive result it can be highlighted that the model always returns the same number of elements at the output as the words in the input sentence, property that the *seq2seq* model has not got. We believe that this model needs a greater capacity of computation and a greater time of training to be able to improve the results.

Table 3 shows the sentences used to test the $1 - LSTM$. As you can see, the model has correctly identified the subject and the predicate in sentences 1, 3, 4, 5, 6 and 7, which is 60%. The model fails when the subject is not at the beginning of the sentence, or in occasions, when the sentence has uniquely subject or predicate. This may be due to the fact that during training all the used sentences contains subject and predicate, and this is always found at the beginning.

Finally the $2 - LSTM$ network is tested. The test consists in passing 10 sentences to the network indicating which words make up the subject and the predicate and

[3] https://www.nltk.org/.

Table 2 Sentences used to test the $2 - LSTM$ model. Teal and blue colours are used to indicate the clause subject and predicate respectively

id.	Sentences
1	Pierre Vinken 61 years old will join the board as a nonexecutive director Nov. 29
2	Mr. Vinken is chairman of Elsevier N.V. the Dutch publishing group
3	The monthly sales have been setting records every month since March

Table 3 Sentences used to test the $1 - LSTM$ network. Teal and blue colours are used to indicate the clause subject and predicate respectively. Last column shows the evaluation of the obtained result (True/False)

id.	Sentences	Correct
1	My parents were on holidays	T
2	My friends and I play football	F
3	They read books	T
4	The newspaper article was written by the journalist	T
5	Do your homework	T
6	The dog ran after the cat	T
7	John and his friends ran outside to play	T
8	The United States president spoke to American people	F
9	Mom asked me to do my homework	F
10	Melanie's mother drove her to the doctor'	F

the type of each word. The model identifies the subject and predicate in all the test sentences, and correctly labels 70 of the 76 words, that is, it obtains a success percentage of 92% in the tested examples (Table 4).

4 Conclusions and Future Works

In this paper, three models of deep neural networks have been presented to perform the syntactic analysis of a sentence. In this first approach, three models have been used: $seq2seq$, $1 - LSTM$ and $2 - LSTM$. $seq2seq$ has been trained and subsequently tested with three sentences of the model used in the learning and has not worked correctly in any case. We think it probably needs a better training. $1 - LSTM$ has been tested using a set of 10 sentences. It has detected the subject and the predicate in 60% of cases, failing mainly in the case that the sentence has no subject or this is not at the beginning. $2 - LSTM$ has also been tested with a set of 10 sentences obtaining 100% in subject and predicate detection. In this case it has also been tested the detection of the type of word reaching a 92% success in this experiment. We

Table 4 Sentences used to test the LSTM bilateral network. Teal and blue colours are used to indicate the clause subject and predicate respectively. Red colour is used to indicate when the labelling fails. Last column shows the evaluation of the obtained result (True/False)

id.	Subject	Predicate	Correct
1	My parents PRP NNS	were on holidays VBD IN NNS	T 100%
2	My friends and I PRP$ NNS CC PRP	play football VBP NN	T 100%
3	The newspaper article DT NN NN	was written by the journalist VBD VBN IN DT **CD**	T 87.5%
4		Do your homework VBP PRP$ **NNP**	T 66.6%
5	The dog DT NN	run after the cat VBD IN DT **NNP**	T 83%
6	John and his friends NNP CC PRP$ NNS	ran outside to play football VBD **JJ** TO VB NN	T 100%
7	My friends and I PRP$ NNS CC PRP	play football VBP NN	T 83.3%
8	The United States president DT NNP NNPS NNP	spoke to American people VBD TO NNP **NNP**	T 87.5%
8	Mom NNP	asked me to do my homework VBD PRP TO VB PRP **VB**	T 100%
9	Melanie's mother NNP POS NN	drove her to the doctor VBD PRP TO DT NNP	T 100%
10	The large man in front of me DT JJ NN IN NN IN NNS	blocked my view VBP PRP$ NN	T 100%

consider encouraging results are obtained from these tests. As future works, a more exhaustive set of tests must be carried out using a larger and more heterogeneous data-set to obtain a more reliable result. In addition, it is interesting to perform the full syntactic analysis indicating the function of each group of words within the sentence. To do this first a large set of data must be obtained from correctly analysed sentences indicating the type of each word and group of words that will allow learning and testing the method.

References

1. Deng, Y., Bao, F., Kong, Y., Ren, Z., Dai, Q.: Deep direct reinforcement learning for financial signal representation and trading. IEEE Trans. Neural Netw. Learn. Syst. **28**(3), 653–664 (2017)
2. Lakhani, P., Sundaram, B.: Deep learning at chest radiography: automated classification of pulmonary tuberculosis by using convolutional neural networks. Radiology **284**(2), 574–582 (2017)
3. Bhave, A., Sharma, M., Janghel, R.R.: Music generation using deep learning. In: Wang, J., Reddy, G., Prasad, V., Reddy, V. (eds.) Soft Computing and Signal Processing. Advances in Intelligent Systems and Computing, vol. 898. Springer, Singapore (2019)
4. Goodfellow, I., Bengio, Y., Courville, A.: Deep Learning. MIT Press, Cambridge (2016). http://www.deeplearningbook.org
5. Levine, R.D.: Syntactic Analysis: an HPSG-Based Approach. Cambridge Textbooks in Linguistics. Cambridge University Press, Cambridge (2017)
6. Di, W., Bhardwaj, A., Wei, J.: Deep Learning Essentials: Your Hands-on Guide to the Fundamentals of Deep Learning and Neural Network Modeling. Packt Publishing, Birmingham (2018)

Estimating Remaining Time of Business Processes with Structural Attributes of the Traces

Ahmad Aburomman, Manuel Lama, and Alberto Bugarín-Diz

Abstract In this paper, we deal with one of the challenges in process mining enhancement: prediction of remaining times in a business process, which is a critical task for many organisations. Our approach consists of (i) defining a number of attributes on the business logs that capture structural information from the traces, (ii) extending the well-known annotated transition system model to annotate its states with the attributes values and (iii) applying linear regression for predicting the remaining time of the process for each state using the attributes values. Experiments with ten well-known real-life datasets show that our approach outperforms the baseline model defined in Van der Aalst et al. (Inf Syst 36(2):450–475, 2011) in the three metrics considered.

Keywords Business process management · Remaining time estimation · Business intelligence

1 Introduction

Information systems managing massive and variant business processes usually store all the transactions data into the form of event logs [9]. In business process management, the usage of event logs is not limited to storing the data generated by the business processes. Moreover it can also be used to make these processes visible,

This research was funded by the Spanish Ministry for Science, Innovation and Universities (grant TIN2017-84796-C2-1-R) and the Galician Ministry of Education, University and Professional Training (grants ED431C 2018/29 and "accreditation 2016–2019, ED431G/08"). All grants were co-funded by the European Regional Development Fund (ERDF/FEDER program).

A. Aburomman · M. Lama · A. Bugarín-Diz (✉)
Centro Singular de Investigación en Tecnoloxías Intelixentes (CiTIUS), Universidade de Santiago de Compostela, 15782 A Coruña, Spain
e-mail: alberto.bugarin.diz@usc.es

M. Lama
e-mail: manuel.lama@usc.es

M. E. Cornejo et al. (eds.), *Computational Intelligence and Mathematics for Tackling Complex Problems 2*, Studies in Computational Intelligence 955,
https://doi.org/10.1007/978-3-030-88817-6_6

since hidden processes can be discovered by applying process mining techniques [3, 4]. Event logs could have information concerning, for example, resources, timestamps, or case data. This historical information can be used to build predictive models which can be used to let running process instances learn from the previous records. By extracting the timestamp from the event log and extend it in the process model, it is possible, for instance, to measure the waiting times between process's activities. These can be used to let running process instances learn from the historical records. By extracting the timestamp from the event log and extend it in the process model, it is possible, for instance, to measure the waiting times between process's activities.

One of the most imperative difficulties is predicting the remaining processing time of running cases [1, 2, 4, 8], defined as the required time for a process to be complete. Its accurate estimation during a process run-time is an issue which has raised recently as one of the challenges in business processes management [8] since in many applications it is critical both from the customer's and the organisations point of view. For example, bank customers need to know how long it will take for their loan application, to be reviewed and checked then get the decision to accepting or declining.

Different works in the literature have addressed the problem of prediction time of business process. The model in [3, 6, 7] proposed a methodology that predicts the remaining time of the finished process, by defining and constructing an annotated transition system (ATS) as a support model and attaching a predictive temporal data at each state in the ATS. Nevertheless, the prediction results in these models are modest, mostly because the use a low number of attributes for building the regression models to estimate the remaining time. Because of this, the models are not able to capture enough information for the training model, thus, getting low accuracy of prediction.

In this work, we present a new vector-based and ATS-based approach that considers structural features or attributes related to the process execution such as frequencies, repetitions, cycles, etc. Each ATS state is characterised with a set of vectors whose components are the values of these attributes for each partial trace that fits this state. Based on these vector and on the remaining times of the traces related to them, a linear regression-based predictor has been built for each state.

2 A New Prediction Model for Remaining Time Estimation

In this section, firstly, we will present the most relevant elements of an ATS [7], since our model proposes an extension of this.

Definition 1 (Event) [7]: An event e is described by a unique identifier and it is characterised by its properties, such as its identifier, time-stamp, and the activity which is executed in the time-stamp.

Definition 2 (Trace, Event Log) [7]: A trace T is a sequence of ordered events $\{e_1, e_2, \ldots, e_M\}$. In Table 1 we show two examples of traces involving, respectively, activities A, B (Trace 1) and A, B, C (Trace 2). An event log is a set of traces.

Table 1 Two examples of traces. The superscript in each activity indicates its timestamp (ending time)

Trace no.	Trace	Activities involved
1	$<A^3 A^{18} B^{24} A^{30}>$	A, B
2	$<A^7 B^{10} B^{15} C^{22} C^{31}>$	A, B, C

Definition 3 (Partial trace, State) [7]: A partial trace PT is any continuous part of a trace T that contains one or more events in sequence. For each (partial) trace, three state representations are defined [7]: Set, Multi-Set, and Sequence. In this paper, we focus on the *Set* representation as the basis for our model. In Set representation, each partial trace PT has associated a state which is labelled through the activities in PT and where no repetition of activities is considered (no matter its execution order).

Definition 4 (Transition System, TS) [7]: A transition system TS is a triplet (S, PT, TR), where S is the state space, PT is a set of partial traces, and TR is a transition relation which describes how the system moves from one state to another state. The TS model has different forms depending on the state representation in which it is based on.

Definition 5 (Annotated Transition System, ATS) [7]: Let L be an event log, and TS a transition system based on the Set representation. For a particular measurement function M, an annotation A is assigned to each state of the TS using the measurement M.

Using these definitions we will propose in what follows an extension of the ATS model that includes a set of new structural features which are extracted from traces and capture relevant information about the traces which have an impact on the remaining time estimations.

2.1 Definition of Attributes

To extend the model with predictive information, we consider new attributes extracted from the event log. We define a number of attributes that include useful information about the log, these attributes can have details about each trace's structure and will be later used as predictors in a regression model, in order to provide a good model for remaining time prediction in a running process.

Definition 6 ($Occurrence$): We define the occurrence of an activity A as the number of times A occurs in the partial trace.

Definition 7 ($Cycle$): We define the $Cycle$ of an activity A as the maximum number of times A is repeated in sequence in the partial trace.

Table 2 Example of the attributes value for the trace 1 in Table 1 $<A^3 A^{18} B^{24} A^{30}>$

Attributes	Description	Representation	Att. Value
Occ(A)	How many time A occurs	AABA	3
Occ(B)	How many time B occurs	AABA	1
Cyc(A)	The maximum continuous repeat of A	AABA	1
Cyc(B)	The maximum continuous repeat of B	AABA	0
Pos(A)	The position of last occurrence of A	AABA	4
Pos(B)	The position of last occurrence of B	AABA	3
Dis(A)	The distance between the current activity and the same last activity of A	AABA	1
Dis(B)	The distance between the current activity and the same last activity of B	AABA	0
Dup(A,A)	Duple of AA	AABA	1
Dup(A,B)	Duple of AB	AABA	1
Dup(B,A)	Duple of BA	AABA	1
Dup(B,B)	Duple of BB	AABA	0
Change	Change of Activities	AABA	2
Single	Number of single Activities	AABA	1
Elt	Elapsed time of the current Activity	AABA	30

Definition 8 $(Position)$: Let PT be a partial trace. We define $Position$ of an activity A as the last happening of A in PT. It represents the last index of A happened in PT.

Definition 9 $(Distance)$: Let A be an activity in a partial trace PT. $Distance(A)$ is the index difference distance between the last time A occurs in PT and the previous one (backwards).

Definition 10 $(Duple)$: Let PT be a partial trace and state S the state associated to PT. For all the pairs of activities A_i and A_j in S, we define $Duple(A_i,A_j)$ as the number of times that the sequence $A_i A_j$ happens in PT.

Definition 11 $(Change)$: We define $Change$ as the number of times that an activity is different from the previous one, from the beginning of a trace until its last activity.

Definition 12 $(Single)$: We define $Single$ as the amount of activities that are not repeated in cycles from the beginning of partial trace.

Definition 13 $(ElapsedTime)$: We define *Elapsed Time* as the time passed since the beginning of a partial trace until its last activity.

In Table 2 we show an example of these attributes calculated trace 1 in Table 1.

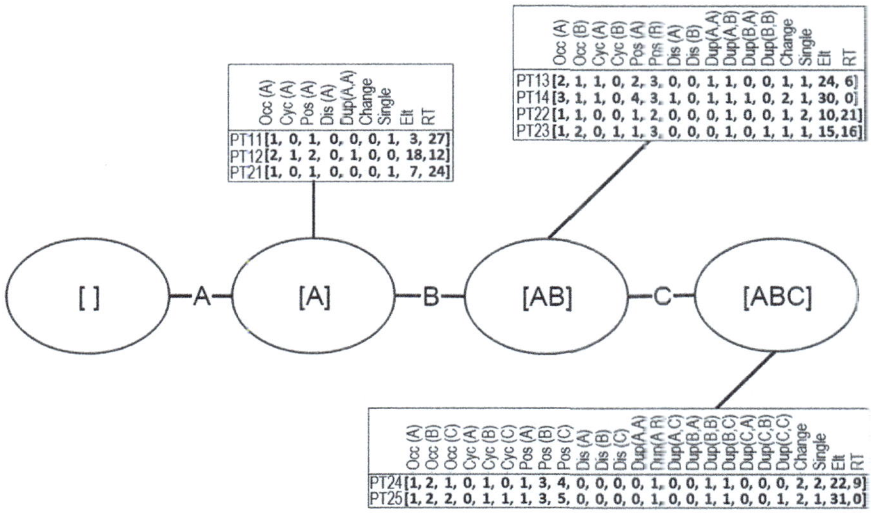

Fig. 1 Extended Annotated Transition System with lists of elements (attributes values) integrated. The traces considered here are the ones indicated in Table 1

Table 3 List of Partial Traces belongs to traces shown in Table 1, where T1 is the first trace and T2 is the second trace in Table 1

	T1	T2
Partial Trace	PT11: $< A^3 >$ PT12: $<A^3A^{18}>$ PT13: $<A^3A^{18}B^{24}>$ PT14: $<A^3A^{18}B^{24}A^{30}>$	PT21: $<A^7>$ PT22: $<A^7B^{10}>$ PT23: $<A^7B^{10}B^{15}>$ PT24: $<A^7B^{10}B^{15}C^{22}>$ PT25: $<A^7B^{10}B^{15}C^{22}C^{31}>$

2.2 *Extended Annotated Transition System*

Previously, we introduced the definition of ATS model, as well as the definitions of the attributes we consider to be extracted from each trace to build our model. In this section, we integrate these attributes into an Extended Annotated Transition System (EATS).

Definition 14 (List): Let S be a state of a transition system ATS and $\{AttR_i, i{=}1,2,...,N\}$ be the set of attributes defined in Sect. 2.1. For all partial traces j associated to S, we define element E_j as the set of their attributes values and the remaining time of trace j, i.e., E_j=[value of $AttR_1$, ..., value of $AttR_N$, RT_j]. Then, we define List(S):= $\{E_1,...,E_P\}$ as the set of elements that annotate state S. Note that each state is annotated with an associated list which includes N elements, where each element contains a set of attributes values evaluated on each of the partial traces represented by S. On the contrary, in [7] authors only annotate each state with a single attribute. In

Table 4 Comparison between our approach and the model in [7] (MAE in days)

		BPIC12	BPIC13	BPIC15_1	BPIC15_2	BPIC15_3	BPIC15_4	BPIC15_5	BPIC17	Hospital	Traffic	Aver.
Accur.	our app.	**0.65**	**0.78**	**0.90**	**0.91**	**0.92**	**0.93**	**0.94**	**0.51**	**0.66**	**0.75**	**0.79**
	[7]	0.39	0.41	0.44	0.48	0.47	0.55	0.56	0.17	0.35	0.50	0.43
MAE	our app.	**1.52**	**1.73**	**3.58**	**8.69**	**1.19**	**3.87**	**4.66**	**8.75**	**19.01**	**31.78**	**8.48**
	[7]	5.69	6.45	42.94	89.68	19.76	20.06	46.83	6.28	52.33	148.36	43.84
MAPE	our app.	**0.01**	2.85	**1.54**	**2.47**	**0.22**	**2.34**	**0.43**	**0.25**	7.59	**0.00**	**1.77**
	[7]	0.24	**2.79**	5.66	9.91	2.67	2.91	6.25	0.40	**7.00**	0.01	3.78

Fig. 1 we show an example of building and annotating the EATS which corresponds to two sample traces shown in Table 1. For instance, state $S = [A]$ is the state associated to the following three partial traces: PT11:=$<A^3>$, and PT12:= $<A^3 A^{18}>$ (from $T1$), and PT21:= $<A^7>$ (from $T2$) as listed in Table 3. Therefore, it has an associated list of three elements made up with the corresponding eight attribute values defined in Sect. 2.1 and the corresponding remaining time for each partial trace (respectively 27, 12, 24).

Once the EATS is built and annotated, linear regression is applied to the list of each state taking the attributes values as independent variables and the remaining time as dependent variable. In this way we obtain an expression for each state that will allow us to estimate the remaining time for any new trace represented by the same state.

3 Experimental Validation and Conclusions

We have validated our model using ten real-life event log of BPI Challenges 2012, 2013, 2015, 2017 and Hospital Bill, and Traffic Fine [5]. Logs in these datasets come from very different fields of applications, such as administrative and financial processes, billing of medical services, road traffic fines management, ... and are considered a de-facto benchmark which is used for Research Challenges in the business process management area. We compared our approach to the baseline ATS model described in [7]. To evaluate the results of our prediction model by calculating the Accuracy of the prediction quality. Moreover, we also used Mean Absolute Error (MAE), and Mean Absolute Percentage Error (MAPE) to measure the error between real remaining time and the predicted remaining time. In Table 4, we can see that in average our model performs better in all ten real-life event logs for all the three metrics considered. In more detail, we can see that in 28 out of the 30 cases (93.3%) our model performs better, being only outperformed by [7] only in two of the MAPE results. Average differences in MAE are 35.36 days, with values ranging from 4.17 to 116.58 days, in all cases favourable to our method. Therefore, we can conclude that our approach produces better results than [7] in a consistent way for a variety of application areas.

Current and future work includes improving the model further by applying regression to data partitions obtained with informative criteria.

References

1. Bolt, A., Sepúlveda, M.: Process remaining time prediction using query catalogs. Lect. Notes Bus. Inf. Process. **171**(1), 54–65 (2014). https://doi.org/10.1007/978-3-319-06257-0_5
2. De Leoni, M., van der Aalst, W.M.P.: Data-aware process mining: discovering decisions in processes using alignments. In: Proceedings of the 28th Annual ACM Symposium on Applied Computing - SAC 13 p. 1454 (2013). https://doi.org/10.1145/2480362.248063

3. Pandey, S., Nepal, S., Chen, S.: A test-bed for the evaluation of business process prediction techniques. In: 2011 7th International Conference on Collaborative Computing: Networking, Applications and Worksharing (CollaborateCom), pp. 382–391. https://doi.org/10.4108/icst.collaboratecom.2011.247129
4. Polato, M., Sperduti, A., Burattin, A., De Leoni, M.: Data-aware remaining time prediction of business process instances. In: Proceedings of the International Joint Conference on Neural Networks, pp. 816–823 (2014). https://doi.org/10.1109/IJCNN.2014.6889360
5. Real-life Event Logs, https://data.4tu.nl/repository/collection:event_logs_real. Last accessed 30th May 2019
6. Van der Aalst, W., Pesic, M., Minseok, S.: Beyond process mining: from the past to present and future. In: Advanced Information Systems, pp. 1–15 (2010). https://doi.org/10.1007/978-3-642-13094-6_5
7. Van der Aalst, W., Schonenberg, M., Song, M.: Time prediction based on process mining. Inf. Syst. **36**(2), 450–475 (2011). https://doi.org/10.1016/j.is.2010.09.001
8. Van der Aalst, W., Adriansyah, A., De Medeiros, A.K.A.: Process mining manifesto. In: Lecture Notes in Business Information Processing 99 LNBIP(PART 1), pp. 169–194 (2012). https://doi.org/10.1016/j.is.2011.10.006
9. Van Dongen, B.F, Crooy, R.A., van der Aalst. W.: Cycle time prediction: when will this case finally be finished?. In: International Conference on the Move to Meaningful Internet Systems (OTM 2008), Lecture Notes in CS, 5331, pp. 319–336 (2008). https://doi.org/10.1007/978-3-540-88871-0_22

Convolutional Neural Networks in the Ovarian Cancer Detection

Piotr A. Kowalski, Jakub Błoniarz, and Łukasz Chmura

Abstract The main goal of the work is to create an artificial neural network capable of recognizing tumor cell groups in histopathological slides. Here, the Convolution Neural Network was used in the course of research on the detection of areas that include ovarian cancer cells. Theoretical and practical investigations included the issues of both preparing data for learning and selecting the appropriate architecture of the artificial neural network. The obtained results show that neural networks can be used as a competitive method for designating areas in which pathological cells of cancer origin are present. An additional goal of the work is to obtain superior results without losing the quality of the input image as one of the major limitations to this is the memory and computing capabilities of the computer on which the tests are carried out.

Keywords Artificial intelligence · Neural network · Deep learning · Convolutional neural network · Image analysis · Histopathological analysis · Ovarian cancer

P. A. Kowalski (✉) · J. Błoniarz
Faculty of Physics and Applied Computer Science, AGH University of Science and Technology, al. Mickiewicza 30, 30-059 Cracow, Poland
e-mail: pkowal@agh.edu.pl; pakowal@ibspan.waw.pl

J. Błoniarz
e-mail: jakub.bloniarz@gmail.com

P. A. Kowalski
Systems Research Institute, Polish Academy of Sciences, ul. Newelska 6, 01-447 Warsaw, Poland

Ł. Chmura
Department of Pathomorphology, Jagiellonian University Medical College, Cracow, Poland
e-mail: lukasz.chmura@icloud.com

M. E. Cornejo et al. (eds.), *Computational Intelligence and Mathematics for Tackling Complex Problems 2*, Studies in Computational Intelligence 955,
https://doi.org/10.1007/978-3-030-88817-6_7

1 Introduction

Nowadays, many life activities are commonly automated. This is due to the pace of life—the need to do more in a shorter time, and because the new adaptations and skills have requirements that are often unable to be met. Here, computers, data analysis [16], machine learning, artificial intelligence [19] and in particular computational intelligence procedures [11, 13] come to our aid. The possibilities of artificial intelligence are almost limit in exploring their potential is our imagination.

The current state of knowledge regarding the use of artificial intelligence is very wide. The first application, however, was in 1959, when the first fully functional MADALINE system was created in an American laboratory in order to reduce echo in telephone conversations. Nowadays, these are very complicated programs based on data on the human genome or protein structure and AI has found wide application in the field of medicine, such as in screening for tuberculosis [4].

Systems modelled on the human brain are not entirely perfect as it is still possible to improve them. Moreover, scientists and engineers have been looking for additional places to implement them. In this work, an attempt was made to combine these models with medicine and image processing. The current state of understanding allows us to put into practice complex algorithms and to solve these only with the help of a computer and from the knowledge acquired by it. Perhaps it will not be long before we need to learn new methods of solving problems, if only so as to be able to pass them to the virtual world.

The main purpose of this article is to show an artificial neural network capable of recognizing groups of cancer cells occurring in histopathological photos. Figures obtained from a research group dealing with a group of female gonadal ovarian cell tumors provide the basis for the creation of the simulation. To create an application that is supportive of the histopathologists' work, the Python programming language was used, and a program for pre-processing of figures was written in it. It is thought that the obtained results will be able to show that through using intelligent methods, an easier path can be found to the designated goal by machine outlining of areas in which there are pathological cells of cancer.

The paper is organised as follows. In the first section (Sect. 2), some general information on histopathological analysis is provided. In Sect. 3, the reader encounters generalized information about data preparation and Convolutional Neural Networks (CNN). The results of numerical verification with discussion are included in Sect. 4. Finally, conclusions are provided regarding the proposed procedure.

2 Histopathological Preliminaries

To realize the purpose of this work, one should first explain what biopsy diagnosis is and what it is, why it is so important, and why it takes so much time.

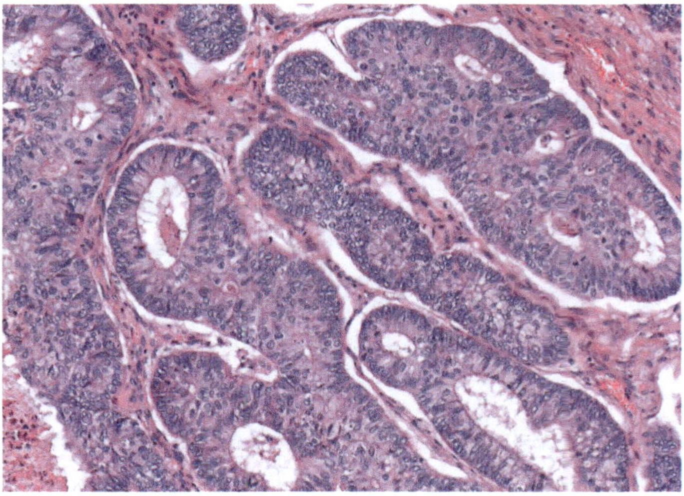

Fig. 1 A histopathological image of mucosal tumor and healthy cells

The diagnosis of cancer itself begins with the tumor being traced within the body using imaging techniques such as MRI and CT [14]. Thus, a potential accumulation of cancer cells is made evident. However, such information is not enough, and confirmation is done through thick-needle and thin-needle biopsy. To do so, using a specialized tool, a cylindrical fragment of the tumor is taken, and then processed and sliced into a very thin layer of cells resembling a sheet of paper [24]. The essence of this diagnosis is the detection of the type of cancer cells that are being dealt with. In this study, cells were used that were derived from ovarian tissue in which the main types of cancers are: serous, non-differentiated, mucous, endometrial and mesonephroid. In this work, only two types: serous and mucous are considered.

Figure 1 depicts a sample figure of a histopathological slice with cells of mucosal tumors.

The type of cancer determines the name of its internal function, height growth or resistance to treatment, hence the large weight of the result of the diagnosis. It will show what available operations we should do and take steps to cure diseases. This diagnosis is aimed at cutting the cancer, and may be attached to a microscope by histopathologists. It takes a long time because each sample requires careful examination to accurately create the type of cancer, because one type found does not mean that there is nothing more that can mean a mixed type.

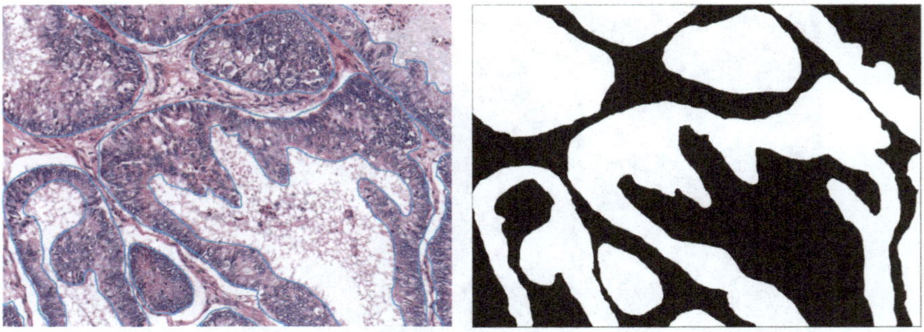

Fig. 2 Histopathological image (*left*) and matrix indicating the presence of tumor cells (*right*)

3 Data Pre-procesing and Convolutional Neural Network

The initial quality of the slides was high, and the images themselves were characterized by a high resolution of 1698 × 1242. It was decided to employ an algorithm that will, from each of the 26 slides, cut out 2,600 smaller images that will be used to learn the network. This group was divided into training and test elements. The newly created photos were 100 × 200 in size pixels, and the groups are divided in a ratio of 9:1. At the same time, 20% of the total was separated in the training group in order to be used for validation.

An important aspect was the simultaneous distinction of the element from the histopathological portrait of the matrix indicating the presence of tumor cells (Fig. 2).

As the name suggests, the artificial neural network algorithms are modelled to a greater or lesser extent on the features of the actual nerve cell. Their mathematical models can most often be considered in the context of a universal approximator of multi-variable functions. The general scheme of a single neuron cell consists of an adder that sums the signals at its input, multiplying them by the appropriate weights and an activation function that processes the signal from the adder output non-linearly. An important aspect in creating the model is the decision on how to teach neural networks. In this work, due to the data set used, the supervised learning approach will be used.

Deep learning is a field of artificial neural networks used to reproduce highly complex processes that require a lot of machine work, such as: face recognition, reading handwritten text. This model is much more complicated than the shallow network model. In its interior there is much more than one hidden layer. It is currently assuming that a network with more than two hidden layers can be called a deep neural network. The number of hidden layers is defined by the longest path that information must transfer from the network's input to its output.

The imaging data prepared in way described in first part of this section, was then analysed by the neural network [17]. The choice fell on employing the Convolutional Neural Network [15]. This type of neural network is especially useful for working

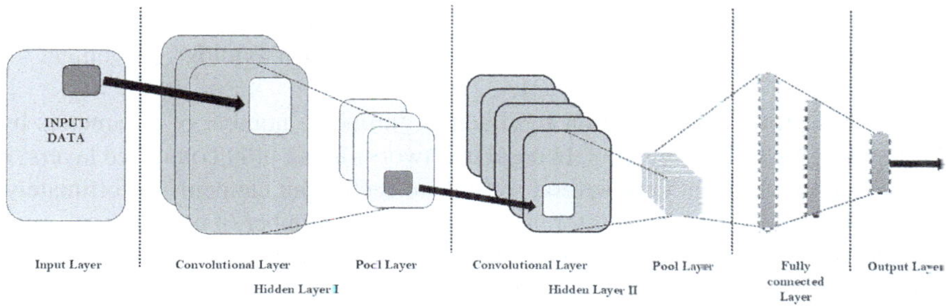

Fig. 3 Topology of convolutional neural network

with images and for analysing video sequences [6], as well as for modelling data in the form of a time series [3, 15]. It solves problems related to the large size of the input image by its scaling in individual layers [2]. This network, however, is not without flaws, but these can be used to advantage. The first and largest is the learning time. The initial stage in creating a multi-thread system of convolutional values is the appropriate selection of their parameters. This cannot be done without assessing the quality of their operation first. For this purpose, data sets are used that are much smaller than those employed for proper learning. In this way, the best configuration of the model can be created.

An exemplary topological diagram describing CNN components is shown in Fig. 3. In the convolutional layers, simple features are extracted in the initial layers of the network, such as edges, spots, and in the following layers it goes to a higher, more abstract level of features. This is because the convolution layers contain various forms of filters that extrude other characteristics of the input.

In addition to the aforementioned convolution layer associated with the filtering operation, another component of this network is the pooling layer, which is used to progressively reduce the spatial size to reducing the number of features and, consequently, reducing the complexity of the network. Most often in convolution networks, we use the MaxPool layer that shifts 2×2 filters through the entire matrix, generating the largest value from the filter window and saves it to the next layer of the network.

One can also use special type of transfer function to eliminate the problem of the disappearing gradient, which results from the fact that the derivatives are in the range [0, 1], so often multiplying these values leads to obtain very small numbers leading to smaller and smaller weight changes [20]. This problem occurs in particular in layers distant from the network exit. To address this problem CNN networks can use the ReLU transformation defined as: $f(x) = max(0, x)$ as a transfer function instead of a tanh or logistic function [7].

In the second part of the CNN network, we can see a fragment that after the flattening operation resembles the classic structure of the so-called shallow neural network. The purpose of flattening is to transform the data collected in the form of abstract multidimensional matrices containing higher-order features into classic vectors. These, in turn, are the entrance to the second part of the neural network. The structure of this part of the neural network may adopt a topology convenient for regression [1, 25] or classification [21, 22].

The application of convolution significantly limits the number of parameters by sharing weights and using filters. In regular networks using fully connected layers, it is necessary to assign at least several weights to each input element. Unfortunately, this solution does not scale well for large data sets; the number of necessary neurons explodes when increasing the size of the considered data. Using CNN networks increases the size of the input without the need to expand the architecture. Thanks to the significant reduction in the number of parameters compared to previous solutions, it is possible to run such a network on any device; hence, the networks are universal and have a large spectrum of applications.

Of course, the convolution-type neural network clearly has more differences, but because of the limit, only the most important details are shown, more information can be found in the CNN literature [18, 23].

In order to properly select the architecture of the network structure [10], it was necessary to carry out several trials, initially concerning the size of the artificial neural network [12]. Due to the fact that the number of images in the sample increased significantly by dividing them, only part of the training group need be used so as to avoid long waiting time for the uploading and processing of images, and to minimize learning time. In order to investigate the best number of convolutional wastes, 7,000 images were used from a total of 60,840. The types of transfer functions in individual layers were examined in a similar way. For optimization of the model, the ADAM algorithm [9] was used. This is characterized by high efficiency and it does not change the learning gradient of individual values or the speed of their learning.

4 Numerical Results

Model learning and evaluation was done with the help of Tensorflow. This Google Inc. programming library under a free license, Apache 2.0 allows you to create DL models quickly and efficiently without going into low-level details, which speeds up the transfer of ideas from the experiment stage from production. This work uses the Tensorflow API (application programming interface), made available in the python programming language.

The application was tried out in numerous tests in order to optimize the outcome and because of the sensitivity of the parameters that were controllable. Thus, firstly, it was established that CNN will consists of five layers of convolution and two fully connected layers at the output. In addition, the convolutional layers are to have 50, 75, 100, 150, 200 filters, respectively, and 625 dense layers and 20,000 neurons. All layers have an *elu* transfer function, with the exception of the second fully connected layer, the output of which passes through a sigmoidal function. The ADAM algorithm [9]] for learning task is thus used, and as a loss function, *binary crossentropy* was employed [8]. Herein, 64 images were utilized on a stand-alone basis. This parameter

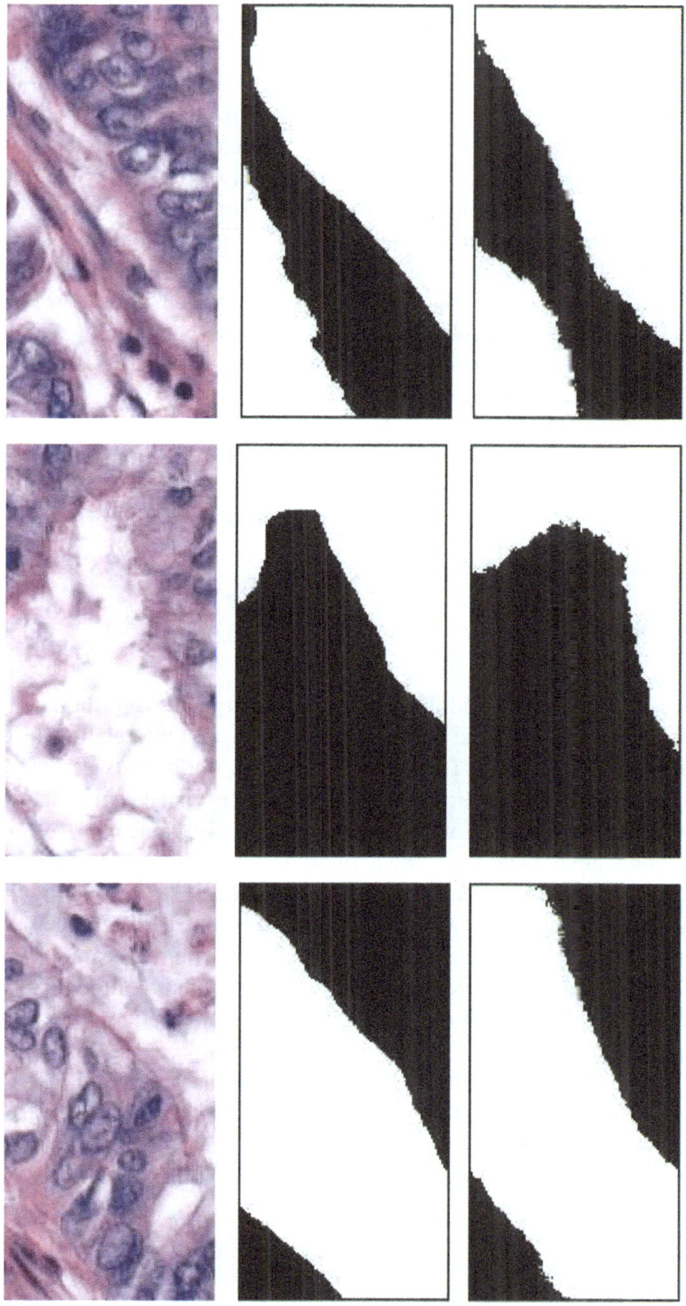

Fig. 4 Simulation effects: from the left: input image, manually marked cells, results of the convolutional neural network

is limited by graphics card's capability. In addition, the number of epochs chosen was 10.

For the model that was implemented, a high value of learning accuracy and the highest of all accuracy of validation were obtained. These values were respectively 0.85 and 0.69 (in the eighth epoch, 0.67). In addition, the error made during the learning process dropped to 0.33 and for validation, to 0.6 (for the 8th epoch, 0.7). It is also worth noting that the parameters connected with the learning process behaved in a predictable manner, however, validation was subject to continuous fluctuations.

In order to evaluate the final version of the learning procedure, a test was prepared consisting of assessing a network of data with which it had no previous contact. As a result of the test, the error on the test data was 0.40 and the accuracy for the test data was 0.82. An eye-to-eye comparison of manually marked images and the results of the neural network can be seen in Fig. 4.

In these images, three examples of cancer cell detection are noticeable. In each case, the image on the left shows the original histopathological portraiture, the middle image is the image containing the identification of diseased cells made by the histopathologist, while the last one is the result of the neural algorithm.

5 Conclusions

From the results of the test and the results that emerged during the learning of the network, it can be seen that the network has learned well the relationships between cancer cells and healthy cells. There is a large compatibility between model images and results created in the application. Moreover, their visual representation is even more convincing that the network has been successfully trained.

An additional advantage is that the network is able to detect errors created during the creation of response matrices, such as the detection of stroma among the labelled cancer cells. Inaccuracy in a small part of cases may be due to differences in the colors of the sample labelling. Normalizing the color scale could possibly eliminate this problem, but this was not necessary to obtain the correctness of the results achieved at the level of 82%. What is more, this result is comparatively high to that obtained in the work [5].

In the future, this study will be extended to include additional types of cancer cells, but this is will require additional samples, and much more computing power. In current conditions, with the processor and graphics card of standard quality, it is not possible to go beyond 60,000 learning samples—which still stretches the capabilities of ordinary computers.

References

1. Babu, G.S., Zhao, P., Li, X.-L.: Deep convolutional neural network based regression approach for estimation of remaining useful life. In: International Conference on Database Systems for Advanced Applications, pp. 214–228. Springer (2016)
2. Christ, P.F., Elshaer, M.E.A., Ettlinger, F., Tatavarty, S., Bickel, M., Bilic, P., Rempfler, M., Armbruster, M., Hofmann, F., D'Anastasi, M., et al.: Automatic liver and lesion segmentation in CT using cascaded fully convolutional neural networks and 3D conditional random fields. In: International Conference on Medical Image Computing and Computer-Assisted Intervention, pp. 415–423. Springer (2016)
3. Cui, Z., Chen, W., Chen, Y.: Multi-scale convolutional neural networks for time series classification (2016). arXiv:1603.06995
4. Darsey, J.A., Griffin, W.O., Joginipelli, S., Melapu, V.K.: Architecture and Biological Applications of Artificial Neural Networks: a Tuberculosis Perspective, pp. 269–283. Springer, New York (2015)
5. Deniz, E., Şengür, A., Kadiroğlu, Zehra, Guo, Y., Bajaj, V., Budak, Ü.: Transfer learning based histopathologic image classification for breast cancer detection. Health Inf. Sci. Syst. **6**(1), 18 (2018)
6. Fan, Y., Lu, X., Li, D., Liu, Y.: Video-based emotion recognition using CNN-RNN and c3D hybrid networks. In: Proceedings of the 18th ACM International Conference on Multimodal Interaction, pp. 445–450. ACM (2016)
7. Hinton, G.E., Salakhutdinov, R.R.: Reducing the dimensionality of data with neural networks. Science **313**(5786), 504–507 (2006)
8. Janocha, K., Czarnecki, W.M.: On loss functions for deep neural networks in classification. CoRR (2017). arXiv:1702.05659
9. Kingma, D.P., Ba, J.: Adam: a method for stochastic optimization. CoRR (2015). arXiv:1412.6980
10. Kowalski, P.A., Kusy, M.: Determining significance of input neurons for probabilistic neural network by sensitivity analysis procedure. Comput. Intell. **34**(3), 895–916 (2018)
11. Kowalski, P.A., Kusy, M.: Sensitivity analysis for probabilistic neural network structure reduction. IEEE Trans. Neural Netw. Learn. Syst. **29**(5), 1919–1932 (2018)
12. Kowalski, P.A., Wadas, K.: Triggering Probabilistic Neural Networks with Flower Pollination Algorithm, pp. 107–113. Springer International Publishing, Cham (2020)
13. Kowalski, P.A., Łukasik, S., Kulczycki, P.: Methods of collective intelligence in exploratory data analysis: a research survey. In: International Conference on Computer Networks and Communication Technology (CNCT 2016). Atlantis Press (2016/12)
14. Kowalski, P.A., Kamiński, J., Łukasik, S., Swiebocka-Wiek, J., Gołuńska, D., Tarasiuk, J., Kulczycki, P.: Application of the Flower Pollination Algorithm in the Analysis of Micro-CT Scans, pp. 1–9. Springer International Publishing, Cham (2019)
15. Kowalski, P.A., Sapała, K., Warchałowski, W.: Pm10 forecasting through applying convolution neural network techniques. Int. J. Environ. Impacts 1–12 (2019)
16. Kulczycki, P., Charytanowicz, M., Kowalski, P.A., Lukasik, S.: Identification of atypical (rare) elements - a conditional, distribution-free approach. IMA J. Math. Control Inf. **35**, 923–937 (2018)
17. Kusy, M., Kowalski, P.A.: Weighted probabilistic neural network. Inf. Sci. **430–431**, 65–76 (2018)
18. Lavin, A., Gray, S.: Fast algorithms for convolutional neural networks. In: Proceedings of the IEEE Conference on Computer Vision and Pattern Recognition, pp. 4013–4021 (2016)
19. Łukasik, S., Lalik, K., Sarna, P., Kowalski, P.A., Charytanowicz, M., Kulczycki, P.: Efficient astronomical data condensation using approximate nearest neighbors. Int. J. Appl. Math. Comput. Sci. 1–9 (2019)
20. Nair, V., Hinton, G.E.: Rectified linear units improve restricted Boltzmann machines. In: Proceedings of the 27th International Conference on Machine Learning (ICML-10), pp. 807–814 (2010)

21. Rastegari, M., Ordonez, V., Redmon, J., Farhadi, A.: XNOR-Net: ImageNet classification using binary convolutional neural networks. In: European Conference on Computer Vision, pp. 525–542. Springer (2016)
22. Shen, W., Zhou, M., Yang, F., Yang, C., Tian, J.: Multi-scale convolutional neural networks for lung nodule classification. In: International Conference on Information Processing in Medical Imaging, pp. 588–599. Springer (2015)
23. Simard, P.Y., Steinkraus, D., Platt, J.C., et al.: Best practices for convolutional neural networks applied to visual document analysis. In: Icdar, vol. 3 (2003)
24. Slaoui, M., Fiette, L.: Histopathology procedures: from tissue sampling to histopathological evaluation. Methods in Mol. Biol. (Clifton, N.J.) **691**, 69–82 (2011)
25. Szegedy, C., Toshev, A., Erhan, D.: Deep neural networks for object detection. In: Advances in Neural Information Processing Systems, pp. 2553–2561 (2013)

Analyzing the Performance of TSP Solver Methods

Boldizsár Tüű-Szabó, Péter Földesi, and László T. Kóczy

Abstract In this paper we analyze the efficiency of three TSP solver methods: the best-performing exact Concorde algorithm, the state-of-the-art inexact Helsgaun's Lin–Kernighan heuristic and our Discrete Bacterial Memetic Evolutionary Algorithm (DBMEA). In our analysis the run time predictability was also taken into account, not only the tour quality and the run time properties. Three models (polynomial, exponential, square-root exponential) were fitted to the mean run times of VLSI (Very-large-scale integration) instances up to 20,000 nodes. The DBMEA produces the highest (close to 1) R^2-values for each model. The Concorde algorithm shows very low run time predictability.

Keywords Traveling salesman problem · Concorde · Lin–Kernighan · DBMEA

1 Introduction

The Traveling Salesman Problem (TSP) is a famous and widely studied NP-hard combinatorial optimization problem. It is an important theoretical and applied problem with many application areas (logistics, transportation, planning, microchip manufacturing etc.).

B. Tüű-Szabó (✉) · L. T. Kóczy
Department of Information Technology, Széchenyi István University, Győr, Hungary
e-mail: tuu.szabo.boldizsar@sze.hu

L. T. Kóczy
e-mail: koczy@sze.hu

P. Földesi
Department of Logistics, Széchenyi István University, Győr, Hungary
e-mail: foldesi@sze.hu

L. T. Kóczy
Department of Telecommunications and Media Informatics, Budapest University of Technology and Economics, Budapest, Hungary

M. E. Cornejo et al. (eds.), *Computational Intelligence and Mathematics for Tackling Complex Problems 2*, Studies in Computational Intelligence 955,
https://doi.org/10.1007/978-3-030-88817-6_8

The problem involves a salesman who visits each place exactly once starting the journey from the headquarters of the company, and then he returns to the starting point. The task is to find the route which results in the minimum distance.

Concorde is currently the most efficient exact TSP solver. It is a complex branch and cut algorithm combined with heuristic mechanisms. With Concorde the optimal solution for all 110 TSPLIB instances were determined up to 85,900 cities [1].

The Lin–Kernighan heuristic was proposed by Lin and Kernighan in 1973 [5]. It was able to find the optimal solutions for most instances up to 100 nodes. This local search procedure is the generalization of the k-opt local search, the algorithm repeatedly performs exchanges on the current tour. It uses sequential k-opt moves and non-sequential double bridge moves for tour improvement. An effective implementation of the Lin–Kernighan heuristic was developed by Helsgaun in 2000 [2]. He modified some of the heuristic rules used in the original Lin–Kernighan heuristic for restricting and directing the search. This modifications result in a very efficient algorithm which is currently the most efficient inexact TSP solver algorithm.

The DBMEA algorithm is a population based evolutionary algorithm which combines the discretized bacterial evolutionary algorithm with local search techniques [4]. In each iteration DBMEA uses two operators to improve the population, the bacterial mutation and the gene transfer. In the bacterial mutation operation the bacteria try to improve itself with mutation, in the gene transfer operation the fittest bacteria transfer information to the less fit bacteria in order to improve them. The DBMEA was tested on discrete NP-hard optimization problems, on the TSP and on its variants (TSP with Time Windows [4], Minimum Latency Problem [7], Time Dependent TSP [8]). In the case of all examined optimization problems the DBMEA showed its efficiency: it produced optimal or near-optimal solutions within reasonable amount of time. In the case of the Minimum Latency Problem DBMEA outperformed even the state-of-the-art methods in the literature.

Only some papers have been dedicated to statistically investigating the performance of the TSP solver algorithms. Plotting the run time of the tested instances Applegate et al. claimed that the Concorde may scale exponentially [1]. Hoos and Stützle investigated the run time behavior of the Concorde algorithm on random uniform Euclidean (RUE) TSP instances [3]. The root-exponential scaling model was the most accurate for run times over sets of RUE instances. Mu et al. studied the empirical performance of EAX (Edge assembly crossover) and Lin–Kernighan heuristic on RUE instances up to 4500 nodes [6]. The analysis indicated that the scaling for Lin–Kernighan heuristic lies somewhere between the polynomial and the square-root exponential model.

2 Computational Results

In our experiments VLSI instances were examined up to 20,000 nodes from the VLSI TSP dataset (http://www.math.uwaterloo.ca/tsp/vlsi/). The distances between the nodes were calculated with real values instead of integer values, so the Concorde and the Helsgaun's Lin–Kernighan heurisic were modified. The tests were carried out on an Intel Pentium Dual CPU T2390 1,86 GHz 2 GB RAM workstation. The instances up to 5000 nodes were tested 10 times, the bigger instances were run 5 times.

The Concorde and the Lin–Kernighan heuristic were tested with their default parameters.

2.1 Comparison of Tour Lengths Found by the TSP Solver Algorithms

The tour lengths (best and average values) were compared found by the three examined methods (Table 1). For the instances with more than 2000 nodes (except from pds2566) the optimal real valued tour are not known because the Concorde was not able to solve them within two CPU days. Comparing the results the Helsgaun's

Table 1 Comparison of tour lengths for VLSI instances

Instance	DBMEA		Helsgaun's Lin–Kernighan		Concorde	Gap DBMEA [%]
	Best value	Average value	Best value	Average value	Optimal value	
xql662	2 550.84	2550.84	2 550.84	2 550.92	2 550.84	0
dkg813	3 243.41	3 243.41	3 243.41	3 243.68	3 243.41	0
dka1376	4 738.61	4 743.20	4 746.87	4 736.87	4 736.87	0
dcq1389	5 156.43	5 158.06	5 152.06	5 153.08	5 152.69	0.04
dja1436	5 332.09	5 335.93	5 328.21	5 329.01	5 328.21	0.07
icw1483	4 467.22	4 472.12	4 465.88	4 465.94	4 465.88	0.07
rbv1583	5 446.09	5 449.99	5 444.67	5 444.91	5 444.67	0.03
rby1599	5 589.13	5 592.02	5 584.16	5 585.20	5 584.16	0.03
dea2382	8 154.34	8 160.93	8 146.94	8 148.87	–	0.09
pds2566	7 774.91	7 781.57	7 766.71	7 766.89	7 766.71	0.1
bch2762	8 376.18	8 380.63	8 368.56	8 370.33	–	0.13
fdp3256	10 104.36	10 108.69	10 091.29	10 092.15	–	0.13
dkc3938	12 715.99	12 723.46	12 704.22	12 704.31	–	0.09
xqd4966	15 554.91	15 560.49	15 530.82	15 531.00	–	0.16
fqm5087	13 230.21	13 237.87	13 193.20	13 194,00	–	0.33
xsc6880	21 965.24	21 981.22	21 901.80	21 902.82	–	0.36
bnd7168	22 193.6	22 198.00	22 126.25	22 127.98	–	0.30
dga9698	28 273.84	28 276.25	28 187.15	28 189.48	–	0.31
frh19289	57 062.65	57 083.45	56 649.11	56 655.91	–	0.73

Lin–Kernighan heuristic produced the best tours: for the instances where the optimal solution is known it found them, and where it is not known it found lower tour length than the DBMEA.

The DBMEA also resulted in high quality tours: for the two instances under 1000 nodes it found the optimal solution, for the bigger instances it produced near-optimal solutions within 0.73% to the best known solutions.

2.2 *Fitting Curves to the Run Times of the Algorithms*

For each TSP solver method the following three models were fitted to the mean run times of the tested VLSI instances:

- polynomial: $f[a, b](n) = a \cdot n^b$
- exponential: $f[a, b](n) = a \cdot b^n$
- square-root exponential: $f[a, b](n) = a \cdot b^{\sqrt{n}}$

A curve which minimizes the root mean square error (RMSE) was fitted to the mean values of the run times. For the Concorde the pds2566 was the biggest instance used for fitting because it failed to solve bigger instances within 2 CPU days.

Table 2 shows the parameters of the fitted curves of each algorithm. The Lin–Kernighan heuristic resulted in the lowest *b* parameters. It indicates that it will be the fastest algorithm on very large instances.

The RMSE and R^2-values were calculated for each fitting model to determine the accuracy of fitting (Table 3). In the case of Concorde algorithm each model shows poor fitting properties, the R^2-values are smaller than 0.4. It indicates that the run times of the VLSI instances are not predictable for the Concorde algorithm. The Lin–Kernighan heuristic shows better predictability than the Concorde algorithm:

Table 2 Parameters of the fitting models

		a	b
Concorde	Polynomial	0.00326	2.12018
	Exponential	2 996.51	1.00114
	Square-root exponential	375.1996	1.10373
Helsgaun's Lin–Kernighan	Polynomial	0.00034	1.81815
	Exponential	693.3861	1.00018
	Square-root exponential	139.8174	1.03706
DBMEA	Polynomial	0.00190	1.98134
	Exponential	16 247.52	1.00018
	Square-root exponential	2 981.33	1.03883

Table 3 The RMSE and R^2-value of the fitting models

		RMSE	R^2-value
Concorde	Polynomial	18 314.6	0.38
	Exponential	18 391.9	0.3748
	Square-root exponential	18 332.2	0.3788
Helsgaun's Lin-Kernighan	Polynomial	1547.7	0.9268
	Exponential	1535.4	0.9280
	Square-root exponential	1483.9	0.9327
DBMEA	Polynomial	7 726.17	0.9965
	Exponential	20 618.07	0.9753
	Square-root exponential	12 335.27	0.9911

its models have lower RMSE and R^2-values. For the Lin–Kernighan heuristic the square-root exponential model was the most accurate. The best fitting were produced by the DBMEA algorithm for each model, the R^2-values closer to 1. It means that the run time of an instances is more predictable for DBMEA than for the Lin–Kernighan heuristic. For DBMEA the polynomial model was the best-fitting model resulting in the lowest RMSE and highest R^2-value. Figures 1, 2 show the curve of the fitting models for the Lin–Kernighan heuristic and the DBMEA.

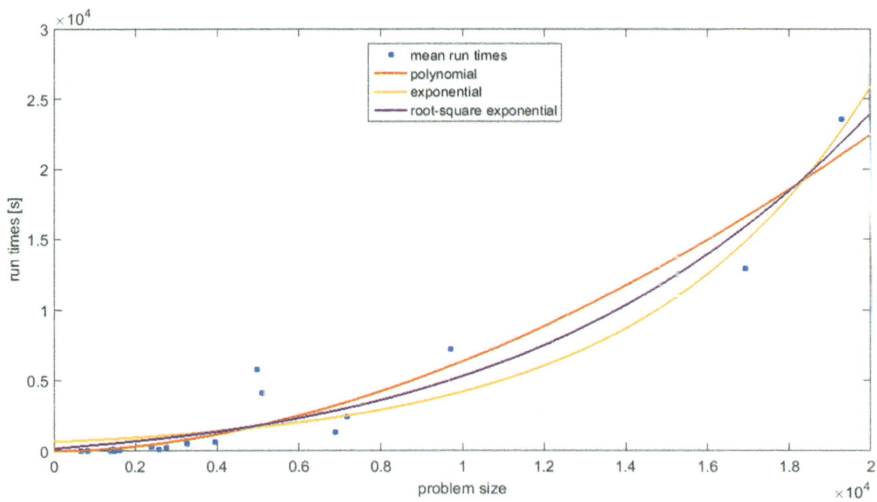

Fig. 1 Fitting models of the Lin–Kernighan heuristic

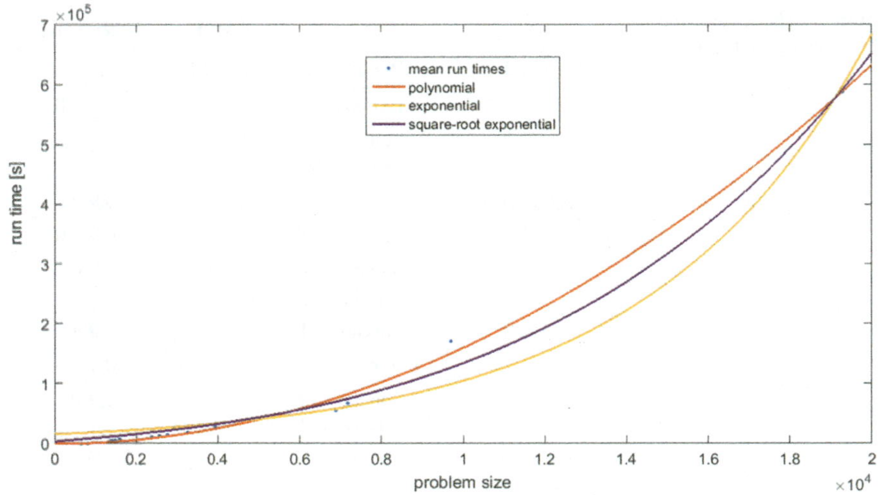

Fig. 2 Fitting models of the DBMEA

3 Conclusion

In this paper three algorithms (Concorde, Lin–Kernighan, DBMEA) were tested on VLSI benchmark instances up to 20,000 nodes. Three models (polynomial, exponential, square-root exponential) were fitted to the mean run times of the instances.

The Lin–Kernighan heuristic was the fastest method and it produced the lowest tour lengths. For each fitting model the DBMEA produced the highest fitting accuracy with close to 1 R^2-values. It indicates that the run time of the DBMEA is well predictable. The Concorde has poor fitting properties, the R^2-values are lower than 0.4 for each model.

Acknowledgements This work was supported by National Research, Development and Innovation Office (NKFIH) K108405, K124055. Supported by the ÚNKP-18-3 New National Excellence Program of the Ministry of Human Capacities. The work was supported by EFOP-3.6.2-16-2017-00015 HU-MATHS-IN – Intensification of the activ-ity of the Hungarian Industrial Innovation Mathematical Service Network.

References

1. Applegate, D.L., Bixby, R.E., Chvátal, V., Cook, W.J.: The Traveling Salesman Problem: A Computational Study, pp. 489–530. Princeton, Princeton University Press (2006)

2. Helsgaun, K.: An effective implementation of the Lin-Kernighan traveling salesman heuristic. Eur. J. Oper. Res. **216**, 106–130 (2000)
3. Hoos, H.H., Stützle, T.: On the empirical scaling of run-time for finding optimal solutions to the travelling salesman problem. Eur. J. Oper. Res. **238**, 87–94 (2014)
4. Kóczy, L.T., Földesi, P., Tüű-Szabó, B.: Enhanced discrete bacterial memetic evolutionary algorithm—An efficacious metaheuristic for the traveling salesman optimization. Inf. Sci. **460**, 389–400 (2017)
5. Lin, S., Kernighan, W.: An effective heuristic algorithm for the traveling-salesman problem. Oper. Res. **21**(2), 498–516 (1973)
6. Mu, Z., Dubois-Lacoste, J., Hoos, H.H., Stützle, T.: On the empirical scaling of running time for finding optimal solutions to the TSP. J. Heuristics **24**(6), 879–898 (2018)
7. Tüű-Szabó, B., Földesi, P., Kóczy, L.T.: A population based metaheuristic for the minimum latency problem. Trends in Mathematics and Computational Intelligence, pp. 113–121. Springer, Berlin (2019)
8. Tüű-Szabó, B., Földesi, P., Kóczy, L.T.: Discrete bacterial memetic evolutionary algorithm for the time dependent traveling salesman problem. In: International Conference on Information Processing and Management of Uncertainty in Knowledge-Based Systems, IPMU 2018, Cádiz, Spain, pp. 523–533. Springer (2018)

A Fuzzy Model to Aggregate Performance Indicators in Sports

Francisco P. Romero, Eusebio Angulo, Jesus Serrano-Guerrero, and Jose A. Olivas

Abstract There is currently a growing interest in evaluating the performance of individual sports teams. These measurement frameworks are based on the opinion of various experts who do not always agree with the importance of the selected indicators. In this work, it is proposed to use methods of treatment of subjectivity and ambiguity based on fuzzy logic to obtain a selection of correctly assessed indicators. As a preliminary case of study, it has been decided to study handball players due to the scarcity of literature on the subject.

Keywords Aggregation operators · Linguistic model · Player performance evaluation

1 Introduction

Over the past few years, we have been witnessing an increasing interest in evaluating the performance of sports teams. Any attempt to assess the process and success of a sports team requires some objective method for valuing players. Sports statistics offer a reasonable and recognizable approach to this issue. Nevertheless, the problem is which statistics should be used. There are some popular measures like goals or assists, but they fail to capture some critical characteristics.

To know which variables influence the performance of a player together with the relative contribution of these variables is especially important for coaches and managers of sports teams. In handball, although the academic and professional attention to quantitative analysis of data has exponentially grown past years, there is no much work on evaluating a player in a game by game scenario [1]. For example, [2] presents

F. P. Romero (✉) · J. Serrano-Guerrero · J. A. Olivas
Department of Information Systems and Technologies, University of Castilla La Mancha, Ciudad Real, Spain
e-mail: franciscop.romero@uclm.es

E. Angulo
Department of Matemathics, University of Castilla La Mancha, Ciudad Real, Spain

M. E. Cornejo et al. (eds.), *Computational Intelligence and Mathematics for Tackling Complex Problems 2*, Studies in Computational Intelligence 955,
https://doi.org/10.1007/978-3-030-88817-6_9

an analysis of the performance of Spanish handball players according to their level revealed that anthropometric characteristics of the top-elite players differ from those of the elite players, mainly in their upper limbs. On the other hand, [3] analyze different performance factors of jump throws in team handball. Their main results are to show that the measured variables can reveal essential performance differences between leagues and playing positions and, taken together, explain most of the variance in ball release height. The work of [4] specifies the elements of team-handball performance based on scientific studies and practical experience, and to convey perspectives for practical implication. Consequently, there is no tool for understanding why players vary their performance from one game to another, nor to know about which variables are affecting performance.

Consensus on how relevant are a selection of performance indicators is lacking and contributes to a wide diversity of approaches. This multiplicity can add cost, impair the ability to focus on the most salient indicators and raise concerns regarding the validity of methods, usefulness, and trust in the obtained results [5]. The complexity of this decision-making problem increases in the form of subjectivity and vagueness. Fuzzy logic is one of the artificial intelligence technology tools to handle ambiguity and uncertainties. The use of a fuzzy approach provides us approximate reasoning methods that can handle the inherent subjectivity. The fuzzy approach has been explored to deal with similar phenomena in decision making.

This paper addresses the problem of identifying variables which may affect the performance of a handball player from one game to another by a group of experts. We propose a fuzzy model to analyze the opinion of several experts and to create definitions by aggregating the specifications of the linguistic labels given by them. Thus, the paper presents a new application that has never been attempted with fuzzy sets, and, according to our preliminary results, produce at least comparable performance with other approaches which use machine learning to automate the process of weighing performance features.

This contribution is set out as follows. Section 2 presents the foundations of our proposals. In Sect. 3, we describe the consensus processes based on fuzzy aggregation functions. Section 4 explains the case of study and shows and discuss the corresponding results. Finally, we present the conclusions and future work in Sect. 5.

2 Background

Let us know formally describe our problem. We assume that there is a group of experts $E_1, E_2, \ldots, E_n$ providing the evaluation of the relevance of each indicator $I_1, I_2, \ldots, I_m, r_{ij}$ denotes the assessment of the indicators I_i according to the expert E_j and is a linguistic value.

There could be missing data, i.e., expert E_j might not provide his/her definition of the indicator F_i. Therefore, for each indicator I_i we have several valuations denoted K, with $K \leq N$. Our objective is to define each I_i as a consensus of the definitions $d_{i1}, d_{i2}, \ldots, d_{ik}$. We can omit the subscript i when the particular I_i is not essential.

In fuzzy logic, aggregations operators usually combine numerical values in [0, 1] but, there is also some extension to combine fuzzy numbers [6]. Among them, we will consider the Type-1 OWA Operator.

2.1 Type-1 OWA Aggregation Operator

According to Zadeh [7], a linguistic variable is formally represented by a quintuple $\langle L, T(L), U, S, Mi \rangle$ [1] where L is the name of the variable; $T(L)$ is a finite term set of (primary) labels or words (a collection of linguistic values); U is a universe of discourse or base variable; S is the syntactic rule which generates the terms in $T(L)$; and M is a semantic rule which associates with each linguistic value X its meaning $M(X) : U \rightarrow [0, 1]$.

Depending on the nature of the problem, there is not only one variable but more, representing different information sources, and sometimes the associated values have to be combined to get an only result. To achieve this, it is necessary an aggregation process which generates final results taking into account all the individual contributions. Many approaches can be found using different types of aggregation operators [8], with the Ordered Weighted Averaging (OWA) operator proposed by Yager [9] being widely used.

An Ordered Weighted Averaging (OWA) operator of dimension n is defined as a mapping $F : R^n \rightarrow R$ which has an associated weighting vector $W = [w_1, w_2, \ldots, w_n]$ in which $w_j \in [0, 1]$ and $\sum_{j=1}^{n} w_j = 1$ and where $F(a_1, a_2, \ldots, a_n) = \sum_{j=1}^{n} w_j b_j$, with b_j being the jth largest of the collection of the aggregated objects a_i.

Inspired by Yager's OWA operator, and Zadeh's extension principle, the Type -1 OWA operator was defined in [10] to allow the fusion of uncertain information with uncertain weights expressed linguistically and represented as (type-1) fuzzy sets.

Given n linguistic weights $\{W_i\}_{i=1}^{n}$ in the form of type-1 fuzzy sets defined on the domain of discourse [0, 1], an associated type-1 OWA operator of dimension n is a mapping Φ,

$$\begin{aligned} \Phi\colon F(X) \times \cdots \times F(X) &\longrightarrow F(X) \\ (A_1, \ldots, A_n) &\mapsto G \end{aligned}$$

that aggregates type-1 fuzzy sets $\{A_i\}_{i=1}^{n}$ in the following way,

$$\mu_G(y) = \sup_{\sum_{k=1}^{n} \bar{w}_i a_{\sigma(i)} = y w_i \in U, a_i \in X} \left(\mu_{W_1}(w_1) * \cdots * \mu_{W_n}(w_n) * \mu_{A_1}(a_1) * \cdots * \mu_{A_n}(a_n) \right) \quad (1)$$

where $*$ is a t-norm operator, $\bar{w}_i = \frac{w_i}{\sum_{i=1}^{n} w_i}$, and $\sigma : \{1, \ldots, n\} \rightarrow \{1, \ldots, n\}$ a permutation function such that $a_{\sigma(i)}$ is the ith largest element in the set $\{a_1, \ldots, a_n\}$.

A fast implementation to perform these calculations can be seen in [11].

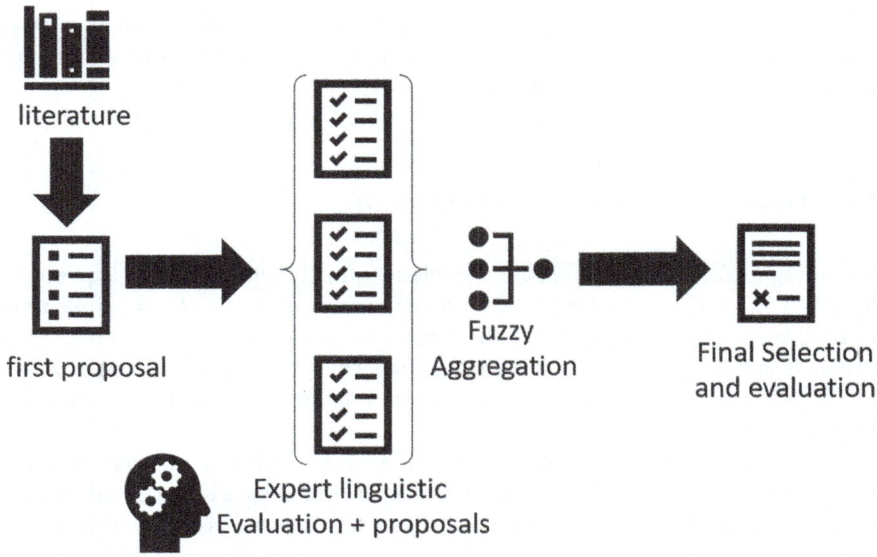

Fig. 1 Graphical abstract of the proposed methodology

3 Methodology

Figure 1 shows a graphical abstract of the proposed methodology.

1. *Baseline/First Proposal*: The starting point must be to know a set of indicators that will be taken into account to model the player performance.
2. *Experts Opinion*: Every expert, depending on his own experiences, must express an opinion regarding the different indicators. In this case, each indicator is evaluated by linguistic labels belonging to the set $S_o = \{S_1, \ldots, S_m\}$, which will be represented by using Triangular Fuzzy Numbers (TFNs).
3. *Aggregation Step*: Once the expert opinions are known, the T1OWA operator is applied to aggregate the set of fuzzy labels representing the evaluations of all indicators. So by using as weights, the collection of fuzzy labels representing the expert preferences S_u. The aggregated output for each opinion will be the fuzzy set representing how relevant each indicator would be to measure the player performance.

4 Case of Study

This section presents the application of the previous proposal to obtain a set of indicators to evaluate handball player performance and its importance according to

Table 1 Indicators baseline

Indicator	Description	Indicator	Description
6mC	6m centre shots	9m	9m shots
Wing	Wing shots	7m%	7mP efficiency
FB	Fast breaks	AS	Assists
R7	Received 7m fouls	TO	Turnover handlings
ST	Steals	BS	Blocked shots
AF	Attack foul	Ex	Exclusions

Table 2 Experts opinions

Indicator	E1	E2	E3	E4	E5	E6	E7	E8	E9	E10	E11	E12	E13	E14	E15
9mMissed	M	M	M	M	M	M	M	M	M	M	M	M	M	M	M
Com7m	M	H	H	M	M	M	H	M	H	M	M	M	L	L	H
Turnover	H	M	M	M	H	L	L	M	M	M	M	H	M	M	M
6mMissed	M	H	H	H	H	H	H	H	H	H	H	H	H	H	H
7mMissed	H	H	H	H	M	H	H	M	M	H	H	H	H	H	H
Exclusion	H	H	H	H	H	H	M	H	M	H	H	H	M	M	H
UDW	H	H	H	M	H	H	H	H	H	H	M	M	H	H	H
Attack foul	M	L	L	M	L	L	L	M	L	M	L	M	L	L	M
WingMissed	M	L	L	L	L	L	M	M	L	L	L	L	M	M	L
6mGoal	M	M	H	M	M	M	H	H	H	H	L	L	M	M	M
7mGoal	M	H	H	H	M	H	H	L	H	H	M	M	M	H	M
Block	M	M	M	L	H	M	L	M	H	M	L	H	M	H	H
DefAct	M	H	H	M	L	H	L	M	H	H	M	H	L	M	H
Received 7m	M	M	M	M	M	M	M	M	H	H	M	M	M	M	M
WingGoal	H	L	L	M	H	H	M	M	H	M	H	H	H	H	H
9mGoal	H	H	H	M	H	M	M	M	H	M	M	M	H	H	H
Assists	H	H	H	H	M	H	H	H	H	H	H	M	H	H	H
FBGoal	M	H	H	M	M	H	H	H	H	H	M	H	H	H	H
FBMissed	M	H	H	M	H	H	H	H	H	H	H	H	H	H	H
Steals	H	M	M	H	H	H	H	H	H	H	H	H	H	H	M

a set of experts. The first stage is to provide a selection of indicators as a baseline. For this purpose, we chose a set of indicators provided by the Spanish Handball Federation used for the evaluation of junior players (Table 1).

A group of 15 experts (handball coaches) evaluates the importance of every indicator, using High (H), Medium (M), Low (L) linguistic labels with the possibility to add a new indicator to the selection (Table 2).

In this case, the aggregated output for each opinion will be the fuzzy set representing how relevant each indicator would be to measure the player performance. Then,

Table 3 Performance indicators final results

Valuation	Indicators
Positive	
High	9mGoal, FBGoal, Steals, Assists
Medium	Defensive actions, 7mGoal, 6mGoal, Received 7m
Low	WingGoal, Block
Negative	
High	Exclusion, 7mMissed, 6mMissed, FBMissed, Uncompleted defensive withdrawal (UDW)
Medium	Com7m, 9mMissed
Low	WingMissed, Attack foul, Turnover

the fuzzy aggregation function is applied. Table 3 shows the results of the importance of each chosen indicator.

These results allow us to define a weighting scheme to evaluate the performance of a handball player in a specific match (see Eq. 2). To compute this weight we first, we assign a value to each action using its statistical impact in the results, then we aggregate this value for each group of indicators (High_positive, High_negative, ...) and, finally we round the final result.

$$\begin{aligned} P_{score} = & \\ & 0.95 * \{High\_positive\} - 0.80 * \{High\_negative\} \\ & 0.85 * \{Medium\_positive\} - 0.70 * \{Medium\_negative\} \\ & 0.65 * \{Low\_positive\} - 0.60 * \{Low\_negative\} \end{aligned} \tag{2}$$

5 Conclusions and Future Work

A fuzzy approach to select those indicator that better represent the contribution of the players to the results is presented in this work. The key idea is to combine the evaluations provided by a group of experts into a single one. We have discussed the application of the Type-1 OWA for this purpose. The case of study, based on the evaluation of handball players by a set of handball coaches, shows many other opportunities to apply this promising proposal to overcome some problems of classic systems.

In the future, it would also be possible to study other scenarios in which the experts are allowed to use already existing linguistic terms, possibly different from the labels used by other experts.

Acknowledgements This work has been partially supported by the Spanish Government under grant MERINET: TIN2016-76843-C4-2-R (AEI/FEDER, UE)

References

1. Saavedra, J.M.: Handball research: state of the art. J. Hum. Kinet. **63**, 5–8 (2018)
2. Ferragut, C., Vila, H., Abraldes, J.A., Manchado, C.: Influence of physical aspects and throwing velocity in opposition situations in top-elite and elite female handball players. J. Hum. Kinet. **63**(1), 23–32 (2018)
3. Jager, J.M., Gail, S., Muller, H.: Evaluation of a test module to measure relevant components of ball release height in jump throws in team handball. J. Hum. Sport Exerc. [S.l.] **12**(1), 17–28 (2017)
4. Wagner, H., Finkenzeller, T., Würth, S., von Duvillard, S.P.: Individual and team performance in team-handball: a review. J. Sports Sci. Med. **13**(4), 808–816 (2014)
5. De Olde, E.M., Moller, H., Marchand, F., McDowell, R.W., MacLeod, C.J., Sautier, M., Bokkers, E.A.: When experts disagree: the need to rethink indicator selection for assessing sustainability of agriculture. Environ. Dev. Sustain. **19**(4), 1327–1342 (2017)
6. Blanco-Mesa, F., León-Castro, E., Merigó, J.M.: A bibliometric analysis of aggregation operators. Appl. Soft Comput. 105488 (2019)
7. Zadeh, L.A.: Is there a need for fuzzy logic? Inf. Sci. **178**(13), 2751–2779 (2008)
8. Yang, Z., Li, J., Huang, L., Shi, Y.: Developing dynamic intuitionistic normal fuzzy aggregation operators for multi-attribute decision-making with time sequence preference. Expert Syst. Appl. **82**, 344–356 (2017)
9. Yager, R.R., Kacprzyk, J. (eds.): The Ordered Weighted Averaging Operators: Theory and Applications. Springer Science & Business Media (2012)
10. Zhou, S.-M., Chiclana, F., John, R.I., Garibaldi, J.M.: Type-1 OWA operators for aggregating uncertain information with uncertain weights induced by type-2 linguistic quantifiers. Fuzzy Sets Syst. **159**(24), 3281–3296 (2008)
11. Zhou, S.-M., Chiclana, F., John, R.I., Garibaldi, J.M.: Alphalevel aggregation: a practical approach to type-1 OWA operation for aggregating uncertain information with applications to breast cancer treatments. IEEE Trans. Commun. **23**(10), 1455–1468 (2011)

Formal Analysis of Solar Power and Weather Data

M. Eugenia Cornejo, Jesús Medina, and Clemente Rubio-Manzano

Abstract Nowadays, the use of renewable energy is a priority for governments around the world since it gives rise to energy saving and environmental sustainability. In the case of photovoltaic systems, it is fundamental, to know how much energy can be generated and how much energy is needed in order to maintain a suitable balance between supply and demand. In this paper, a formal analysis of solar power and weather data is performed in order to study how the energy generated by solar panels depends on sunny days and weather conditions. In particular, a software architecture is proposed which is formed by three modules. The first and second ones allow us to transform open data files to formal contexts. In the third module, two important processes are performed: (i) a characterization of the states of the sky and (ii) an analysis of the weather conditions under which the energy production is optimal.

Keywords Formal concept analysis · Renewable energy · Fuzzy sets

1 Introduction

The present work is an initial study in which fuzzy mathematical tools will be applied with the purpose of improving the processes involved in the management of renewable energies and energy efficiency. Specifically, the company Grupo Energético de Puerto Real S.A., located in Cádiz (Spain), is commended with the management and integral maintenance of the public lighting network of the city. Its main purpose is

M. E. Cornejo (✉) · J. Medina · C. Rubio-Manzano
Department of Mathematics, University of Cádiz, Cádiz, Spain
e-mail: mariaeugenia.cornejo@uca.es

J. Medina
e-mail: jesus.medina@uca.es

C. Rubio-Manzano
e-mail: clrubio@ubiobio.cl

C. Rubio-Manzano
Department of Information Systems, University of the Bío-Bío, Concepción, Chile

M. E. Cornejo et al. (eds.), *Computational Intelligence and Mathematics for Tackling Complex Problems 2*, Studies in Computational Intelligence 955,
https://doi.org/10.1007/978-3-030-88817-6_10

guaranteeing a service to the citizen of the highest quality, minimizing operating costs and impacts to the environment, through energy efficiency. Focusing on this last aspect, we will collaborate with this company analyzing the stored information corresponding to its photovoltaic installations.

With respect to the photovoltaic installations, it is fundamental to know how much solar energy can be generated and how much solar energy is needed in order to maintain a suitable balance between supply and demand. As the energy generated by solar panels depends on sunny days and weather conditions (among other factors), it is desirable to have information that allows us to predict possible failures or low energy peaks. One of the most important and critical aspects, related to manage and control this process and other ones, is the associated degree of uncertainty and unpredictability, which highlights the need of using fuzzy mathematical techniques.

In this paper, by using tools of multi-adjoint formal concept analysis [2, 6, 7], we characterize the states of the sky and we study under which weather conditions an optimal energy production is obtained.

2 Preliminary Concepts

2.1 Multi-adjoint Formal Concept Analysis

Multi-adjoint concept lattices [7] were introduced as a new general approach to formal concept analysis, in which the philosophy of the multi-adjoint paradigm was applied to the formal concept analysis, in order to provide a general framework that could conveniently accommodate different fuzzy approaches given in the literature. Now, we recall the main notions associated with the multi-adjoint concept lattice framework.

Definition 1 A *multi-adjoint frame* $\mathcal{L}$ is a tuple $(L_1, L_2, P, \&_1, \dots, \&_n)$ where $(L_1, \preceq_1)$ and $(L_2, \preceq_2)$ are complete lattices, $(P, \leq)$ is a poset and $(\&_i, \swarrow^i, \nwarrow_i)$ is an adjoint triple with respect to L_1, L_2, P, for all $i \in \{1, \dots, n\}$.

Definition 2 Let $(L_1, L_2, P, \&_1, \dots, \&_n)$ be a multi-adjoint frame, a *context* is a tuple (A, B, R, σ) such that A and B are non-empty sets (usually interpreted as attributes and objects, respectively), R is a P-fuzzy relation $R\colon A \times B \to P$ and $\sigma : A \times B \to \{1, \dots, n\}$ is a mapping which associates any element in $A \times B$ with some particular adjoint triple in the frame.

Once we have fixed a multi-adjoint frame and a context for that frame, the concept-forming operators are denoted as ${}^\uparrow : L_2^B \longrightarrow L_1^A$ and ${}^\downarrow : L_1^A \longrightarrow L_2^B$ and are defined, for all $g \in L_2^B$, $f \in L_1^A$ and $a \in A, b \in B$, as

$$g^\uparrow(a) = \inf\{R(a, b) \swarrow^{\sigma(b)} g(b) \mid b \in B\} \tag{1}$$

$$f^\downarrow(b) = \inf\{R(a, b) \nwarrow_{\sigma(b)} f(a) \mid a \in A\} \tag{2}$$

The notion of concept is defined as usual: a *multi-adjoint concept* is a pair $\langle g, f \rangle$ satisfying that $g \in L_2^B$, $f \in L_1^A$ and that $g^\uparrow = f$ and $f^\downarrow = g$.

The definition of concept lattice in this framework is given below. Notice that, the ordering defined below provides $\mathcal{M}$ with the structure of complete lattice.

Definition 3 Given a multi-adjoint frame $(L_1, L_2, P, \&_1, \ldots, \&_n)$ and a context (A, B, R, σ), a *multi-adjoint concept lattice* is the set

$$\mathcal{M} = \{\langle g, f \rangle \mid g \in L_2^B, f \in L_1^A \text{ and } g^\uparrow = f, f^\downarrow = g\}$$

in which the ordering is defined by $\langle g_1, f_1 \rangle \preceq \langle g_2, f_2 \rangle$ if and only if $g_1 \preceq_2 g_2$ (equivalently $f_2 \preceq_1 f_1$).

After presenting briefly the mathematical framework in which is supported this work, we show a short explanation about Open data and JavaScript Object Notation.

2.2 *Open Data and JavaScript Object Notation*

Open data are data that can be freely used, shared and built-on by anyone and anywhere. The open data systems are usually implemented by using the *internet data exchange and manipulation protocol* called REST (Representational State Transfer) in which requests must be made in an information exchange language, because when exchanging data between a browser and a server, the data can only be text [4]. The open data systems employed in this paper uses the called JSON (JavaScript Object Notation) format which is a lightweight data-interchange format, it is easy to read, write, parse and generate. A JSON file is built on two structures: a collection of name/value pairs; an ordered list of values. For example: `var person = { name: "Clemente", age: 36, city: "Cadiz" }`. More complex objects can be created, for more information see [4].

3 From Open Data to Formal Contexts

A software architecture formed by three modules is proposed in this paper. The first and second ones allow us to transform open data files to formal contexts. In the third module, two important processes are performed by using tools of multi-adjoint formal concept analysis: (i) a characterization of the states of the sky and(ii) an analysis of the weather conditions under which the energy production is optimal. A scheme associated with the software architecture for formal data analysis is shown in Fig. 1.

In this section, we focus on the first and the second modules detailing a general method for loading, integrating and transforming a collection of open data in a formal context. Firstly, solar energy and weather data are obtained via REST from the open

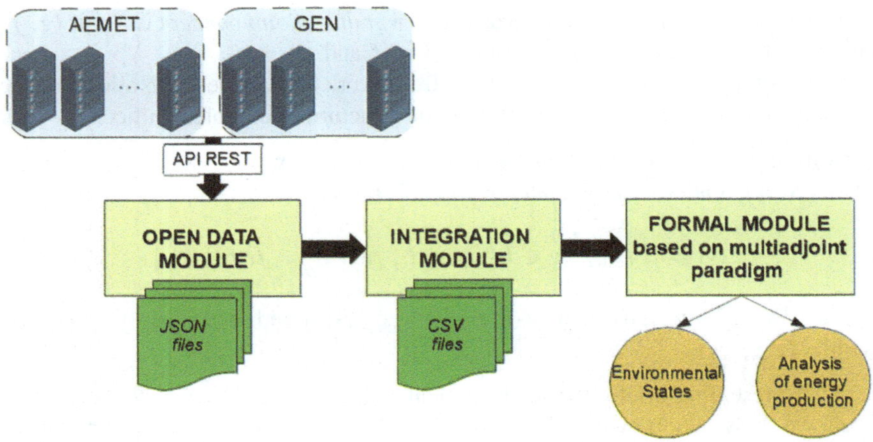

Fig. 1 Scheme associated with the software architecture for formal data analysis

data GEN portal [3] and the open data AEMET portal [1], respectively. Two JSON files are generated and stored in the system, the first one is called *"gen.json"* and it contains the energy production data per hour for a photovoltaic installation during a particular period of time. The structure of the "gen.json" file is as follows:

```
//gen.json
[{                                                    [{
    "cil": "ES0PD1F001",                                  "cil": "ES0PD1F001",
    "serialNumber": "88497884",                           "serialNumber": "88497884",
    "timespan": "2019-05-01T00Z",        ...              "timespan": "2019-05-01T11Z",
    "consumptionKwh": 0,                                  "consumptionKwh": 0,
    "productionKwh": 0                                    "productionKwh": 38
  }]                                                   }]
```

Each object is defined by five attributes: cil (photovoltaic installation key), serial number, timespan, energy consumption of the station in Kwh and the energy production of the station in Kwh.

The second one is called "aemet.json" and it contains the weather prediction data obtained for the next hours. Specifically, at day D, 0 h, we take the prediction for the day D from 14 to 24 h and day $D + 1$ from 0 to 13 h. Each prediction object is defined by using eight attributes: state of the sky (S), rainfall (R), rainfall probability (RP), storms probability (SP), temperature (T), relative humidity (RH) and maximum wind speed (WS). An important feature of the weather prediction data is that there exist several kinds of periods. The periods for the attributes state of the sky, temperature and wind speed are indicated by using a simple hour (e.g. 14, 15, 16, so on). However, the periods for the rainfall and storms probability attributes are indicated by using an interval of hours (e.g. "0208"). In order to homogenize it, intervals of hours will

Date	Hour	Cil	ProductionKwh	S	R	RP	T	RH	SP	WS
2019-05-01	12	P1	43	HighClouds	0	0	24	24	0	34
2019-05-01	13	P1	46	SlightlyCloudy	0	0	26	23	0	43
2019-05-01	14	P1	44	SlightlyCloudy	0	0	26	22	0	42
2019-05-01	15	P1	41	SlightlyCloudy	0	0	27	22	0	42
2019-05-01	16	P1	30	SlightlyCloudy	0	0	26	21	0	42

Fig. 2 Table merging gen.json and aemet.json files to a unique relational file

be transformed in a list of hours, e.g., "0208" will be transformed in the following list {2, 3, 4, 5, 6, 7, 8}.


```
"prediction" : {
"day" : [ { "skystate" : [ { "value" : "16", "period" :  14",
    "description": "CoveredSky" }, ...]
    "rainfall": [ {"value" : "0", "period" : "14"}, ...]
    "RainfallProb": [ { "value" : "", "period" : "0208" }, ]
    "StormProb": [ { "value" : "",""period" : "0208"  }, ...]
    "Temperature": [ { "value" : "15", "period" : "14"  },...]
    "RelativeHumidity": [ { 'value" : "62",  "period" : "14"  },..]
    "WindAndMaxSpeed" : [ { "direction" : [ "O" ], "speed" : [ "28" ],
    "period": "14"  }, ...]}
```


Secondly, an integration phase must be performed in order to homogenize the structure of both JSON files obtained in the previous phase. For that, we are going to employ CSV files since they allows us to represent the data in a scalable and very simple way. Two steps are performed here: (1) "gen.json" and "aemet.json" files are transformed in "gen.csv" and "aemet.csv" files ; (2) "gen.csv" and "aemet.csv" are merged in a unique CSV file which is called "gen_aemet.csv" (see Fig. 2).

Thirdly, a formal context is created from the previously generated CSV file. A formal context is formed, where the objects are given by the day and the hour, for example, the day 2019-05-01 and hour 12 is associated with the object "2019-05-01:12.0". The attributes are given by the production (ProductionKwh) and the rest of meteorological data. In order to obtain the relation of the context, two important steps must be performed here: (i) normalization for each quantitative column by calculating the maximum (M) of their values and dividing each one by this maximum M; (ii) creation of a set of columns for each qualitative column, that is, for each value that appears in the column, a new column is created. For example, from the column S whose values are {HighClouds, SlightlyCloudy, Cloudless, Cloudless, Cloudless, Very Cloudy, CoveredSky}, then seven new columns will be created with their respective values, namely: HighClouds = {1, 0, 0, 0, 0}; SlightlyCloudy = {0, 1, 0, 0, 0}; Cloudless = {0, 0, 1, 1, 1}; Very Cloudy = {0, 0, 0, 0, 0}; CoveredSky = {0, 0, 0, 0, 0}. Finally, Table 1 is created from the data shown in Fig. 2. The two decimal digits are considered as granularity, that is, the values of the relation are in $[0, 1]_{100}$. Moreover, we will consider the following simplification: HC (High

Table 1 Different rows of the relation of the formal context

Objects/Attributes	ProductionKwh	T	R	WS	HC	SLC	C	VC	CS
2019-05-01:12.0	0.89	0.77	0.36	0.42	1	0	0	0	0
2019-05-01:13.0	0.95	0.83	0.35	0.53	0	1	0	0	0
2019-05-01:14.0	0.91	0.83	0.33	0.52	0	0	1	0	0
2019-05-01:15.0	0.85	0.87	0.33	0.52	0	0	1	0	0
2019-05-01:16.0	0.62	0.83	0.32	0.52	0	0	1	0	0

Clouds), SLC (SlightlyCloudy), C (Cloudless), VC (Very Cloudy) and CS (Covered-Sky).

Once we have established the fuzzy relation between the sets of attributes and objects, we can apply some tools developed in multi-adjoint formal concept analysis framework in order to analysis of energy production and weather data.

4 Study of Environmental States

In this section, we will carry out a study of the environmental states, due to their importance in the energy generation. In order to reach this goal, we will compute the concepts generated by the following attributes: high clouds, slightly cloudy, cloudy, cloudless, very cloudy and covered sky. From the obtained concepts, we will find information about the objects and attributes related to the previous states of the sky. Specifically, we will analyze the information collected in May 2019, by Grupo Energético Puerto Real S.A, on the data of energy production in its photovoltaic installation ID:07.

4.1 General Study of Environmental States

First of all, we will study the information collected during the time slot from 0 to 24 h and we compute the concepts generated by each attribute commented above. For example, in the concept generated by cloudless, the values given by the attributes explain the minimum value that these attributes are present when we have a cloudless day. In Fig. 3, we see that the minimum temperature, humidity, and wind speed are 0.45, 0.18 and 0.04, respectively, which correspond, by the normalization that in a cloudless day, to at least a temperature of 14 degrees, a relative humidity of 18% and a wind speed of 3 km/h. Notice that the production does not appear, because at least during one hour of cloudless the installation did not produce energy.

The grey column includes the number of objects in the extension of the concept, that is, the number of objects satisfying these thresholds. Therefore, from the consideration of these concepts, we can characterize the states of the sky according to the

Attribute	Temperature	Relative Humidity	Wind Speed	#Objects
HighClouds	0.48	0.23	0.03	142.0
SlightlyCloudy	0.45	0.21	0.03	146.0
Cloudless	0.45	0.18	0.04	399.0
Cloudy	0.51	0.81	0.13	6.0
CoveredSky	0.54	0.66	0.07	12.0
Very Cloudy	0.51	0.63	0.07	9.0

Fig. 3 Attribute concepts of state of sky generated from 0 to 24 h

Attribute	ProductionKwh	T	RH	WS	RP	#Objects
HighClouds	0.54	0.64	0.36	0.27	0	31.0
SlightlyCloudy	0.33	0.67	0.32	0.23	0	30.0
Cloudless	0.62	0.61	0.27	0.21	0	82.0
CoveredSky	0.5	0.67	0.95	0.41	0.33	3.0
Very Cloudy	0.54	0.67	1.0	0.47	0.33	1.0

Fig. 4 Attribute concepts of state of sky generated from 12 to 16 h

environmental conditions. In the following section we will only consider the hours from 12 to 16, in which we can ensure energy production.

4.2 *Particular Study of Environmental States*

As we previously commented, in order to relate the production with the meteorological and environmental conditions, we will consider the information collected during the time slot from 12 to 16 h, which are the hours with maximum energy generation

In this case, we can also extract interesting information. For example, that the productivity of the installation is obtained on a cloudless day (from 12 to 16), specifically, it produces at least 24 Kw per hour. Moreover, the temperature is not an important factor for the productivity. Furthermore, we have that from 12 to 16 the wind usually is lesser in a cloudless day than in another day. In addition, we can think that the wind is stronger in a cloudy or covered sky day, however, this deduction is only based on 4 cases, as we can see in the #Objects column. In the future, we will add more meteorological variables for obtaining more information.

5 Analysis of Energy Production

In this section we are going to analyze the most interesting fuzzy implications [5, 8, 9] related to the attribute ProductionKwh. Specifically, two fuzzy attribute implications

Rules obtained from 12 to 16 hours	Support	Confidence
ProductionKwh/0.33, Temperature/0.61, RelativeHumidity/0.27, Wind/0.21 ←	1.0	1.0
ProductionKwh/0.62, Cloudless/1.0 ← ProductionKwh/0.33, Temperature/0.61, RelativeHumidity/0.27, Wind/0.21, Cloudless/0.01	0.546	1.0

Fig. 5 Fuzzy attribute implications

will be analyzed. The definition of these implications and variables support and confidence are given in [5]. The first implication in Fig. 5 is a fact, that means that the minimum level of energy production, temperature, relative humidity and wind from 12 to 14 in May is 16 Kwh (0.33), 19 degrees (0.61), 27% and 17 km/h (0.21), approximately. The second one explains that in a Cloudless day the energy production increases to 30 Kwh (0.62), approximately, with total confidence.

6 Conclusions and Future Work

A first approximation for performing formal analysis of energy production and weather data, by using formal concept analysis, has been introduced. A method for translating open data to formal contexts has been detailed. A formal analysis for several attributes and values has been proposed and performed, focusing on the state of the sky and the energy production, and interesting conclusions have been obtained. As a future work, more meteorological variables will be included and a deeper study will be done, showing a comparison with results obtained by other frameworks for data analysis. The application of multi-adjoint concept lattices will be applied to other datasets related to Industry 4.0 and digital forensic.

Acknowledgements The authors acknowledge Dr. Enrique Daneri Vías and Ms. Verónica Carrasco Bernal, from Grupo Energético de Puerto Real, Cádiz, Spain, their valuable collaboration in the preparation of this paper. This paper has been partially supported by the State Research Agency (AEI) and the European Regional Development Fund (FEDER) project TIN2016-76653-P, and Grupo Energético de Puerto Real (GEN). Furthermore, this work has been performed in collaboration with the research group SOMOS (SOftware-MOdelling-Science) funded by the University of Bío-Bío.

References

1. Aemet Open Data (https://opendata.aemet.es/opendata/documentation/swagger-ui.html)
2. Cornejo, M.E., Medina, J., Ramírez-Poussa, E.: Characterizing reducts in multi-adjoint concept lattices. Inf. Sci. **422**, 364–376 (2018)
3. Gen Open Data (http://77.241.112.100:8077/gen-publicservices-web/rest/swagger-ui.html)

4. Introduction to JSON (https://www.json.org/)
5. Liñeiro-Barea, V., Medina, J., Medina-Bulo, I.: Generating fuzzy attribute rules via fuzzy formal concept analysis. In: Kóczy, L.T., Medina, J. (eds.) Interactions Between Computational Intelligence and Mathematics. Studies in Computational Intelligence. pp. 105–119. Springer International Publishing, Cham (2018)
6. Medina, J., Ojeda-Aciego, M.: Multi-adjoint t-concept lattices. Inf. Sci. **180**(5), 712–725 (2010)
7. Medina, J., Ojeda-Aciego, M., Ruiz-Calviño, J.: Formal concept analysis via multi-adjoint concept lattices. Fuzzy Sets Syst. **160**(2), 130–144 (2009)
8. Rodríguez-Lorenzo, E., Bertet, K., Cordero, P., Enciso, M., Mora, A.: Direct-optimal basis computation by means of the fusion of simplification rules. Discret. Appl. Math. **249**, 106–119 (2018). Concept lattices and applications: recent advances and new opportunities
9. Vychodil, V.: Computing sets of graded attribute implications with witnessed non-redundancy. Inf. Sci. **351**, 90–100 (2016)

A New Community Detection Problem Based on Bipolar Fuzzy Measures

Inmaculada Gutiérrez, Daniel Gómez, Javier Castro, and Rosa Espínola

Abstract In social network research, one of the most important analysis is community detection. Fuzzy uncertainty appears clearly when modeling real situations by means of networks. Nevertheless, most of the algorithms used to detect communities in graphs represent them as something crisp. Due to its speed and efficiency, Louvain algorithm is one of the most popular methods used to find clusters in crisp networks. In this study, we propose a modification of it, based on the incorporation of a bipolar fuzzy measure defined over the nodes of the network. Our proposal is based on the use of the Shapley value, which is considered to measure the importance of each node.

Keywords Bipolar fuzzy clustering · Networks · Community detection · Bipolar fuzzy graph

1 Introduction

The main purpose of community detection problems is to find the communities or clusters in networks [5]. This problem has become very popular along the years, so

---

This research has been partially supported by the Government of Spain, Grant Plan Nacional de I+D+i, MTM2015-70550-P, PGC2018-096509-B-I00, and TIN2015-66471-P.

I. Gutiérrez (✉) · D. Gómez · J. Castro · R. Espínola
Faculty of Statistics, Complutense University, Avenida Puerta de Hierro, s/n, 28040 Madrid, Spain
e-mail: inmaguti@ucm.es

D. Gómez
e-mail: dagomez@estad.ucm.es

J. Castro
e-mail: jcastroc@estad.ucm.es

R. Espínola
e-mail: rosaev@estad.ucm.es

M. E. Cornejo et al. (eds.), *Computational Intelligence and Mathematics for Tackling Complex Problems 2*, Studies in Computational Intelligence 955,
https://doi.org/10.1007/978-3-030-88817-6_11

there are numerous effective methods which have been proposed to identify embedded clusters in large networks [24].

Fuzzy measures, that have been deeply studied in the literature, can be understood as a generalization of probabilistic, possibilist or plausibility/belief measures. Let N be a set of nodes, let E be a set of edges and let μ be a fuzzy measure over any subset of nodes. Let us use the notion of extended fuzzy graph, $\widetilde{G} = (N, E, \mu)$, that was proposed in [10]. The triplet $\widetilde{G} = (N, E, \mu)$ can be seen as a natural generalization of fuzzy graphs, in which the only available information is provided by the links of the graph. Then, this work can be seen as a continuation of the analysis developed in [10], where the information provided by a fuzzy measure was incorporated to the classical community detection problem, providing the definition of the community detection problem over extended fuzzy graphs. Nevertheless, the information or capacity associated to the fuzzy measure is not always one-dimensional, so it can be decomposed in a factorial manner based on different characteristics. We follow this idea, and we work under the inspiration of the idea of *bipolarity*, considering as the point of departure the Lacroix and Lavency's approach [13]. This concept is referred to the human mind's characteristic way of dealing with information by simultaneously assigning it both a positive and a negative character [14, 21].This ability to deal with both positive and negative information at the same time, allows our mind to be able to represent and manage complex situations related to interests and emotions, such as ambivalence, indeterminacy or conflict [7, 9]. In consonance with these ideas, in this paper we suggest a modification of Louvain algorithm, which allows to consider some additional information defined by a bipolar fuzzy function when detecting communities. This bipolar modeling will provide us a more realistic representation and processing of human judgments, intentions and preferences.

The remainder of the paper is organized as follows. We introduce the community detection problem over extended fuzzy graphs in Sect. 2. Our proposal to modify the Louvain algorithm so that it allows the search of cohesive groups in a network considering the additional information defined by a bipolar fuzzy measure, is given in Sect. 3. In this section we also provide an illustrative example. Finally, we carry out with some conclusions in Sect. 4.

2 Community Detection Problems over Extended Fuzzy Graphs

Let N be a finite set of nodes, and let E be a collection of edges, i.e., unordered pairs $\{i, j\}$, with $i, j \in N$. Then, a graph or network G is a pair $G = (N, E)$. A graph can also be represented by means of its adjacency matrix, called A. Then, the problem of finding a *good* partition in a graph is known as *Community Detection Problem*. In this paper we will quantify the reliability of a partition, by means of the modularity measure proposed in [17], which we will be denoted as Q. Then, given a partition P of the set of nodes, its modularity $Q(P)$ is defined as:

$$Q(P) = \frac{1}{2|E|} \sum_{ij} \left[A_{ij} - \frac{k_i k_j}{2|E|} \right] \delta_{ij}$$

where k_i represents the degree of node i, and the value of δ_{ij} is 1 if nodes i and j are in the same community, and 0 otherwise.

An extended fuzzy graph can be defined as a triplet $\widetilde{G} = (N, E, \mu)$, where $G = (N, E)$ is a graph, and μ is a fuzzy measure over the set of nodes, i.e., a function $\mu : 2^N \longrightarrow [0, 1]$, where $\mu(\emptyset) = 0$, $\mu(N) = 1$, and for all $A \subseteq B \subseteq N$, $\mu(A) \leq \mu(B)$. Let us note that if $\forall A \subseteq N$, $\mu(A) = 1$, then the graph is crisp.

A new algorithm based on the classical Louvain method was introduced in [10]. It was defined to deal with extended fuzzy graphs. In that paper, the fuzzy measure μ is built based on a matrix that represents the similarity of each pair of nodes. We would like to emphasize that modularity measure has been extended for those cases in which the graph is directed or valued (see [19]) or those in which obtained partition is fuzzy (see [8]). Nevertheless, this measure has not been extended for the case of extended fuzzy graphs, so at the moment, we are not able to quantify if a solution for an extended fuzzy graph is good or not. Even that in [10], we can see that the classical problem changes the final solution in a reasonable way with that class of information.

Notwithstanding the foregoing, let we remark that not only positive information is useful to join nodes as clusters in community detection problems. We can also take advantage of the negative information when determining which nodes should be in the same group. Then, the idea of bipolar fuzzy measures is introduced in this work where we show it is very useful for dealing with that class of information [4, 23].

A bipolar fuzzy measure over the set N, μ^b, can be defined as a function $\mu^b : 2^N \longrightarrow [0, 1]^2$, where, $\forall A \subseteq N$, $\mu^b(A) = (\mu^-(A), \mu^+(A))$, being μ^- and μ^+ fuzzy measures, each of them representing the negative and positive fuzzy evidence associated with the subset A, respectively.

Definition 1 Let $G = (N, E)$ be a graph, and let $\mu^b = \left(\mu^-, \mu^+\right)$ be a bipolar fuzzy measure over the set of nodes N. Then, the triplet $\widetilde{G} = (N, E, \mu^b)$ is denoted as **extended bipolar fuzzy graph**.

3 An Adaptation of Louvain Algorithm for Extended Bipolar Fuzzy Graphs

In this section, we provide a modification of Louvain algorithm, defined in [1], which will allow us to consider additional information (independent of the structure of the graph) when dealing with a community detection problem.

Given a graph $G = (N, E)$, let us suppose that G is connected; otherwise, its connected components can be analyzed separately. Also, let us assume that we have a bipolar fuzzy measure, $\mu^b = (\mu^-, \mu^+)$, which defines a relation among the nodes of the network. Hence, we have an extended bipolar fuzzy graph, $\widetilde{G} = (N, E, \mu^b)$.

In order to deal in an efficient way with the problem of how to adapt the Louvain algorithm for extended bipolar fuzzy graphs, we are going to divide the information into two different parts. On the one hand, we have the adjacency matrix of the graph $G = (N, E)$, denoted as A. It represents the direct connections between nodes. On the other hand, we have the bipolar fuzzy measure μ^b, which represents the capacity of each subset of nodes. Let us emphasize that there could exist two nodes, i, j, directly connected in the graph, i.e. $\exists\ e = \{i, j\} \in E$, but unrelated through μ^b, or the opposite situation, in which there is no edge between a pair of nodes for which μ^b defines a strong relation.

First, let us give some basic notions about Louvain algorithm. Given the graph $G = (N, E)$, whose adjacency matrix is A, the input parameter of Louvain algorithm is A. The main idea of this algorithm is trying to maximize the modularity of a partition, calculating it based on matrix A. Then, we can assume that mentioned matrix is used in two different ways: the first use is intended to take into account the links that are part of the graph, i.e., to consider only pairs of nodes that are connected in G; the second use is intended to maximize the modularity of obtained partitions.

So, we can see Louvain algorithm as an algorithm with two input parameters, $Louvain(matrix_{conexion}, matrix_{maximization})$. Specifically, in the classical version, both input parameters are the same, the adjacency matrix A. Then, we can understand the classical method as $Louvain(A, A)$.

Then, our proposal is quite simple: it consists on the application of the algorithm $Louvain(A, \mathcal{M})$, where $\mathcal{M}$ is some matrix different to A. This modification will not only consider the direct connections of the graph, but will also take into account the information of $\mathcal{M}$ when it comes to maximizing the modularity of the obtained partition. Let us illustrate the new algorithm by means of a pseudocode.

Algorithm 1 Modified Louvain **input**=$(A,\ \mu^b = (\mu^-, \mu^+))$ **output**=P

1: Aggregation of the information of the fuzzy measures μ^- and μ^+ into the matrices $\mathcal{F}^-$ and $\mathcal{F}^+$, respectively:
2: Aggregation of matrices A, $\mathcal{F}^-$ and $\mathcal{F}^+$ into the matrix $\mathcal{M}$, whose modularity will be maximized in the algorithm.
3: Louvain(A, $\mathcal{M}$).

A particular case of the Algorithm 1 will be specified in the following algorithm. Let us assume there are two matrices, $\mathcal{F}^-$ and $\mathcal{F}^+$, obtained from a specific aggregation of the information of the fuzzy measures μ^- and μ^+, respectively (in [10] can be found a concrete case in which fuzzy measures are additive). Then, matrices $\mathcal{F}^-$ and $\mathcal{F}^+$ will represent how each node of a pair is affected by the absence of the other one, depending of their relation.

- The Shapley value measures the importance of the nodes. Then, the minimum [22] of the subtraction between the importance in the presence of all nodes, (Sh_i and

Algorithm 2 Specific Case of Modified Louvain **input**=$\left(A,\quad \mu^b = \left(\mu^-, \mu^+\right)\right)$ **output**=P

1: $\mathcal{F}^-_{ij} = min\{Sh_i(\mu^-) - Sh^j_i(\mu^-), Sh_j(\mu^-) - Sh^i_j(\mu^-)\}$
$\mathcal{F}^+_{ij} = min\{Sh_i(\mu^+) - Sh^j_i(\mu^+), Sh_j(\mu^+) - Sh^i_j(\mu^+)\}$
2: $\mathcal{M} = \alpha\mathcal{A} + (1-\alpha)\left(\frac{1}{2}\mathcal{F}^{-on} + \frac{1}{2}\mathcal{F}^+\right)$ where
$\mathcal{F}^{-on} = \frac{I^- - F^-}{\sum_i^n \sum_j^n \left(I^-_{ij} - F^-_{ij}\right)}$, and $I^-_{ij} = max\{F^-_{ij} | i, j \in N\}, \forall i \neq j; I^-_{ii} = 0$.
3: Louvain$(A, \mathcal{M})$.

Sh_j), and the importance in the presence of all nodes except one, (Sh^j_i and Sh^i_j), indicates the relation between nodes i and j (discrepancy in the case of $\mathcal{F}^-$, or affinity in the case of $\mathcal{F}^+$).

- We want to aggregate the matrices $\mathcal{A}$, $\mathcal{F}^-$ and $\mathcal{F}^+$ by means of a linear combination, in order to obtain the matrix $\mathcal{M}$. As matrices $\mathcal{F}^-$ and $\mathcal{F}^+$ have different meanings, (discrepancy and affinity, respectively), we need to define the matrix $\mathcal{F}^{-on}$, which is normalized and the opposite of $\mathcal{F}^-$. Then, we are able to aggregate the matrices $\mathcal{A}$, $\mathcal{F}^{-on}$ and $\mathcal{F}^+$ with weights α, $\frac{1}{2}(1-\alpha)$ and $\frac{1}{2}(1-\alpha)$, respectively. Let us emphasize that if $\alpha = 1$, the partition problem of the graph will be similar to the classical community detection problem.

Let us note that the calculation of Shapley value has an exponential complexity. Although this calculation problem can be partially solved by means of sampling techniques (see [2, 3] for more details), in the following example, we will only focus on additive fuzzy measures, so that the exact calculation of Shapley value does not imply a high computational cost.

Example 1 Let us assume $\widetilde{G} = \left(N, E, \mu^b\right)$ is an extended bipolar fuzzy graph, whose structure is a chain with 12 nodes as it can be seen in its adjacency matrix (a), and whose bipolar fuzzy measure, $\mu^b = \left(\mu^-, \mu^+\right)$ provides the matrices $\mathcal{F}^-$ and $\mathcal{F}^+$ (matrices (b) and (c) respectively). We will show how our proposal gathers the relations given by μ^b, by applying the Algorithm 2.

$$\frac{1}{11}\begin{pmatrix} 0&1&0&0&0&0&0&0&0&0&0&0\\ 1&0&1&0&0&0&0&0&0&0&0&0\\ 0&1&0&1&0&0&0&0&0&0&0&0\\ 0&0&1&0&1&0&0&0&0&0&0&0\\ 0&0&0&1&0&1&0&0&0&0&0&0\\ 0&0&0&0&1&0&1&0&0&0&0&0\\ 0&0&0&0&0&1&0&1&0&0&0&0\\ 0&0&0&0&0&0&0&1&0&1&0&0\\ 0&0&0&0&0&0&0&1&0&1&0&0\\ 0&0&0&0&0&0&0&0&1&0&1&0\\ 0&0&0&0&0&0&0&0&0&1&0&1\\ 0&0&0&0&0&0&0&0&0&0&1&0 \end{pmatrix}$$

(a) Adjacency Matrix A

$$\frac{1}{12}\begin{pmatrix} 0&0&0&0&0&0&1&0&0&0&0&0\\ 0&0&0&0&0&0&1&0&0&0&0&0\\ 0&0&0&0&0&0&1&0&0&0&0&0\\ 0&0&0&0&0&0&0&0&0&0&0&0\\ 0&0&0&0&0&0&0&0&0&0&0&0\\ 0&0&0&0&0&0&1&1&1&0&0&0\\ 1&1&1&0&0&1&0&0&0&0&0&0\\ 0&0&0&0&0&1&0&0&0&0&0&0\\ 0&0&0&0&0&1&0&0&0&0&0&0\\ 0&0&0&0&0&0&0&0&0&0&0&0\\ 0&0&0&0&0&0&0&0&0&0&0&0\\ 0&0&0&0&0&0&0&0&0&0&0&0 \end{pmatrix}$$

(b) Matrix $\mathcal{F}^-$ given by μ^-.

$$\frac{1}{18}\begin{pmatrix} 0&0&0&0&0&1&0&0&0&0&0&0\\ 0&0&0&0&0&1&0&0&0&0&0&0\\ 0&0&0&0&0&0&0&0&0&0&0&0\\ 0&0&0&0&0&0&0&0&0&0&0&0\\ 0&0&0&0&0&0&0&0&0&0&0&0\\ 1&1&0&0&0&0&0&0&0&0&0&0\\ 0&0&0&0&0&0&0&1&0&1&1&0\\ 0&0&0&0&0&0&1&0&0&1&0&1\\ 0&0&0&0&0&0&0&0&0&0&0&0\\ 0&0&0&0&0&0&1&1&0&0&1&1\\ 0&0&0&0&0&0&1&0&0&1&0&0\\ 0&0&0&0&0&0&0&1&0&1&0&0 \end{pmatrix}$$

(c) Matrix $\mathcal{F}^+$ given by μ^+.

In Table 1, it can be seen that, considering the case of a chain with 12 nodes, the classical version of the algorithm defines 3 clusters with 4 nodes each one. Although

Table 1 Clusters and modularity obtained by means of Louvain or our proposal (Algorithm 2)

	Our proposal (Algorithm 2)				Louvain algorithm			
α	$\mathcal{Q}_A$	$\mathcal{Q}_{\mathcal{M}}$	$\alpha\mathcal{Q}_A + (1-\alpha)\mathcal{Q}_{\mathcal{M}}$	Clusters	$\mathcal{Q}_A$	$\mathcal{Q}_{\mathcal{M}}$	$\alpha\mathcal{Q}_A + (1-\alpha)\mathcal{Q}_{\mathcal{M}}$	Clusters
1	0.4835	−0.0509	0.4835	{{1, 2, 3, 4}, {5, 6, 7, 8}, {9, 10, 11, 12}}	0.4835	−0.0509	0.4835	{{1, 2, 3, 4}, {5, 6, 7, 8}, {9, 10, 11, 12}}
0,9	0.4835	−0.0509	0.4301	{{1, 2, 3, 4}, {5, 6, 7, 8}, {9, 10, 11, 12}}	0.4835	−0.0509	0.4301	{{1, 2, 3, 4}, {5, 6, 7, 8}, {9, 10, 11, 12}}
0,7	0.4091	0.2114	0.3498	{{1, 2, 3, 4, 5, 6}, {7, 8, 9, 10, 11, 12}}	0.4835	−0.0509	0.3232	{{1, 2, 3, 4}, {5, 6, 7, 8}, {9, 10, 11, 12}}
0,5	0.4091	0.2114	0.3103	{{1, 2, 3, 4, 5, 6}, {7, 8, 9, 10, 11, 12}}	0.4835	−0.0509	0.2163	{{1, 2, 3, 4}, {5, 6, 7, 8}, {9, 10, 11, 12}}
0,3	0.4091	0.2114	0.2707	{{1, 2, 3, 4, 5, 6}, {7, 8, 9, 10, 11, 12}}	0.4835	−0.0509	0.1094	{{1, 2, 3, 4}, {5, 6, 7, 8}, {9, 10, 11, 12}}
0,1	0.4091	0.2114	0.2312	{{7, 8, 9, 10, 11, 12}, {1, 2, 3, 4, 5, 6}}	0.4835	−0.0509	0.0025	{{1, 2, 3, 4}, {5, 6, 7, 8}, {9, 10, 11, 12}}

it is a *good* partition when only the topology of the graph is considered, let us note that in this example there is some additional information about the relations among different pairs of nodes, which should be used. Then, it can be assumed that, as far as possible, nodes with a *good* relation should be in the same cluster, and the opposite for pairs of nodes which have a *bad* relation. In matrix (b), we can see that the pairs of nodes $6-7, 6-8$ and $6-9$ have a *bad* relation, but the classical partition keeps them together in the same community. On the other hand, in matrix (c), we can see that some pairs of nodes have a *good* relation, for example pairs $1-6, 2-6, 7-10$ or $8-10$, are in different clusters in the classical partition. Let us note that when some importance is assigned to the additional information, $(\alpha < 0.9)$, and obeying to the topology of the graph, our algorithm is able to break away nodes which have *bad* relations, and to keep together nodes which have *good* relations.

4 Conclusions

The main aim of this work is the proposal of a modification of Louvain algorithm. This modification of the classical algorithm is able to consider some additional information, not inherent to the topology of the network, when it comes to solving community detection problems.

Given a graph, $G = (N, E)$, whose adjacency matrix is called A, we suppose there exists some additional information about the relations among nodes, which is independent of the structure of the graph, defined by a bipolar fuzzy measure $\mu^b = (\mu^-, \mu^+)$. Hence, we have an extended bipolar fuzzy graph, $\widetilde{G} = (N, E, \mu^b)$. Our proposal is to incorporate this additional information to Louvain algorithm in order to obtain communities consistent with all the available information.

For a basic example, we have shown that our algorithm provides a final aggregated modularity better than this obtained with Louvain algorithm. The fuzzy structure relative to the nodes is also gathered with our algorithm.

We would like to emphasize that in this work we are dealing with a community detection problem in which a bipolar fuzzy measure defines some additional information about the relation among the nodes. This possibility comprises a novelty in fuzzy clustering literature.

As further work, we will develop a wider comparison of our method, in order to test statistically the results that are intuited in this study. It is well established that Louvain algorithm performs well with very large networks [6, 15 16, 20], so we will work in the development of computational tests applied in large graphs, comparing our proposal with other clustering methods and considering different measures to quantify the goodness of the results.

Beyond our immediate further research, it will be interesting to apply our idea in more complex context, as, for example, *fuzzy cognitive maps* (FCM). FCM are defined by concepts (nodes) and relationships (weighted arcs), and are usually used as decision support tools. Then, we can see a FCM as a weighted, directed graph in which each concept represents a key factor, state of the modeled system or variable, and weighted and directed edges between concepts indicate cause and effect relationships and the strength [11, 12]. FCM are evolving constructions, but still, represented by similar type of bipolar graphs [18]. Due to this similarity, we will check the goodness of our algorithm in more complex fuzzy graphs.

References

1. Blondel, V., Guillaume, J., Lambiotte, R., Lefevre, E.: Fast unfolding of communities in large networks. J. Stat. Mech.-Theory Exp. **P10008** (2008)
2. Castro, J., Gómez, D., Tejada, J.: Polynomial calculation of the Shapley value based on sampling. Comput. Oper. Res. **36**(5), 1726–1730 (2009)
3. Castro, J., Gómez, D., Molina, E., Tejada, J.: Improving polynomial estimation of the Shapley value by stratified random sampling with optimum allocation. Comput. Oper. Res. **82**, 108–188 (2017)
4. De Tre, G., Zadrozny, S., Matthe, T., Kacprzyk, J., Bronselaer, A.: Dealing with positive and negative query criteria in fuzzy database querying bipolar satisfaction degrees. In: Lecture Notes in Artificial Intelligence. vol. 5822, pp. 593–604. Flexible Query Answering Systems: 8th International Conference, FQAS 2009 (2009)
5. Fortunato, S.: Community detection in graphs. Phys. Rep.-Rev. Sect. Phys. Lett. **486**(3–5), 75–174 (2010)
6. Fortunato, S., Hric, D.: Community detection in networks: a user guide. Phys. Rep.-Rev. Sect. Phys. Lett. **659**, 1–44 (2016)
7. Franco, C., Rodríguez, J., Montero, J.: Building the meaning of preference from logical paired structures. Knowl.-Based Syst. **83**, 32–41 (2015)
8. Gómez, D., Rodríguez, T., Yáñez, J., Montero, J.: A new modularity measure for fuzzy community detection problems based on overlap and grouping functions. Int. J. Approx. Reason. **74**, 88–107 (2016)
9. Grabisch, M., Labreuche, C.: A decade of application of the Choquet and Sugeno integrals in multi-criteria decision aid. Ann. Oper. Res. **175**(1), 247–286 (2010)
10. Gutiérrez, I., Gómez, D., Castro, J., Espínola, R.: A new community detection algorithm based on fuzzy measures. In: Kahraman, C., Cebi, S., Cevik Onar, S., Oztaysi, B., Tolga, A., Sari, I. (eds.) Advances in Intelligent Systems and Computing Series. vol. 1029. Intelligent and Fuzzy Techniques in Big Data Analytics and Decision Making. INFUS 2019. Springer, Cham (2020)
11. Harmati, I., Koczy, L.: On the existence and uniqueness of fixed points of fuzzy set valued sigmoid fuzzy cognitive maps. In: IEEE International Conference on Fuzzy Systems (FUZZ-IEEE) (2018)
12. Hatwagner, M., Yesil, E., Dodurka, M., Papageorgiou, E., Urbas, L., Koczy, L.: Two-stage learning based fuzzy cognitive maps reduction approach. IEEE Trans. Fuzzy Syst. **26**(5), 2938–2952 (2018)
13. Lacroix, M., Lavency, P.: Preferences: putting more knowledge into queries. In: et al, A.I.R. (ed.) VLDB. vol. 1029, pp. 217–225. Thirteenth International Conference on Very Large Data Bases: 1987 13th VLDB (1987)
14. Montero, J., Bustince, H., Franco, C., Rodríguez, J., Gómez, D., Pagola, M., Fernández, J., Barrenechea, E.: Paired structures in knowledge representation. Knowl.-Based Syst. **100**, 50–58 (2016)
15. Newman, M.: Communities, modules and large-scale structure in networks. New J. Phys.**12**(103018) (2010)
16. Newman, M.: Communities, modules and large-scale structure in networks. Phys. Rev. **8**(1), 25–31 (2012)
17. Newman, M., Girvan, M.: Finding and evaluating community structure in networks. Sci. Rep. **69**(2) (2004)

18. Papageorgiou, E., Hatwagner, M., Buruzs, A., Koczy, L.: A concept reduction approach for fuzzy cognitive map models in decision making and management. Neurocomputing **232**, 16–33 (2017)
19. Speidel, L., Takaguchi, T., Masuda, N.: Community detection in directed acyclic graphs. Eur. Phys. J. B **88**(8), 203 (2016)
20. Traag, V., Bruggeman, J.: Community detection in networks with positive and negative links. Phys. Rev. E **80**(3), 036115 (2009)
21. Villarino, G., Gómez, D., Rodríguez, J., Montero, J.: A bipolar knowledge representation model to improve supervised fuzzy classification algorithms. Soft Comput. **22**(15), 5121–5146 (2018)
22. Yager, R., Kacprzyk, J.E.: The Ordered Weighted Averaging Operators. Springer Science & Business Media (2012)
23. Zadrozny, S., Kacprzyk, J.: Bipolar queries: an approach and its various interpretations. In: Carvalho, J., Kaymak, D., Sousa, J. (eds.) International-Fuzzy-Systems-Association World. pp. 1288–1293. Joint International-Fuzzy-Systems-Association World Congress/European-Society-Fuzzy-Logic-and-Technology Conference (2009)
24. Zhao, Z., Zheng, S., Li, C., Sun, J., Chang, L.: A comparative study on community detection methods in complex networks. J. Intell. Fuzzy Syst. **35**(1), 1077–1085 (2018)

Novel Methods of FCM Model Reduction

Miklós F. Hatwágner and László T. Kóczy

Abstract Fuzzy Cognitive Maps are widely applied to support decision making tasks. It is often hard for experts to create the model of a system that provides the required accuracy but simple enough to easily use in practice. In general, it is better to create complex models first, because they can be computationally reduced later until they preserve the required accuracy but become simple enough. Two novel Fuzzy Cognitive Map reduction methods based on K-Means and Fuzzy C-Means clustering are suggested in order to generate simplified models that hopefully mimic the behavior of the original model better than the already existing methods. After the quick overview of the existing techniques found in literature, a simple and a complex model of a real-life problem are reduced to varying degrees with the suggested new methods and with an existing one. The first results of the comparison are published in this paper, too.

Keywords Fuzzy cognitive maps · Model reduction · Clustering

The paper was written with the support of the project titled "Internationalisation, initiatives to establish a new source of researchers and graduates and development of knowledge and technological transfer as instruments of intelligent specialisations at Széchenyi István University" (project number: EFOP-3.6.1-16-2016-00017). M. F. H. acknowledges the financial support of the DE Excellence Program. L. T. K. is supported by NKFIH K108405 and K124055 grants.

M. F. Hatwágner (✉) · L. T. Kóczy
Department of Information Technology, Széchenyi István University, Győr, Hungary
e-mail: miklos.hatwagner@sze.hu

L. T. Kóczy
e-mail: koczy@sze.hu; koczy@tmit.bme.hu

L. T. Kóczy
Department of Telecommunications and Media Informatics, Budapest University of Technology and Economics, Budapest, Hungary

M. E. Cornejo et al. (eds.), *Computational Intelligence and Mathematics for Tackling Complex Problems 2*, Studies in Computational Intelligence 955,
https://doi.org/10.1007/978-3-030-88817-6_12

1 Introduction

Fuzzy Cognitive Map (FCM) is a graph-based technique that is suitable to model complex systems, represent expert knowledge, experiences and using the appropriate methods it is also capable to mimic the reasoning and decision making of humans [12, 13]. This capability predestined this technique to solve decision making problems with it [1, 17, 28], to qualitatively describe the operation of certain systems [23, 24] and support policy making [11, 21] among others [20].

FCM is a bipolar [30], signed fuzzy graph. The key factors, component states or variables are represented by the nodes of the graph, called *concepts* in the FCM literature. The connections among concepts are represented by weighted, directed arcs. The sign of weights expresses the amplifying or weakening effect of a concept on another concept, while the magnitude of weights defines the strength of causal connections. The assigned states of concepts make possible to perform not only static [12] but dynamic [14] analysis of the model, too. Using initial states of concepts and applying a straightforward inference formula, simulations can be performed and "what-if" questions can be answered.

An FCM can be created in two main ways: manually by experts or with machine learning techniques [18, 19, 25]. If historical data are available, machine learning is the proposed method of model construction. If such data are missing, only the first method remains. Unfortunately it is often hard for experts to properly chose the important factors of the system. In case they omit important factors, the resulting model is going to be inaccurate. In the opposite case, the complexity of the model starts to grow rapidly (the possible number of connections is the quadratic function of the number of concepts) making the model cumbersome to use in practice. That is why the authors already proposed [7, 22] to create slightly over-sized models first then reduce them computationally until the balance of accuracy and size becomes acceptable. The idea behind model reduction is to create clusters of concepts and consider these cluster concepts of the reduced model.

This paper is organized as follows. First, a quick overview of FCMs will be provided. The paper continues with the description of the motivating problem. The model of this problem is going to appear later as the over-sized model to be reduced and the results of reduction with different methods will be compared. Obviously, model reduction can be performed in several ways and the paper will summarize some of the possible methods found in the literature. The short description of the already proposed method based on Fuzzy Tolerance Relations (FTR) [7] follows, because our goal is to propose another methods that exceed its results. Later, these novel methods will be introduced and applied on a very simple model and on the motivating, real-life model. The results will be compared and conclusions will be drawn.

1.1 Quick Overview of Fuzzy Cognitive Maps

An FCM can be described formally with a 4-tuple (C, W, A, f), where $C = \{C_1, C_2, \ldots, C_M\}$ is the set of concepts, and the cardinality of the set is M. $W : (C_i, C_j) \rightarrow w_{ij}$ represents the signed weights among concepts. These weights are real numbers from the $[-1, +1]$ interval. According to Kosko's original idea, the maps in this paper never contain self-loops, thus the main diagonal contains zeros. (There are some generalizations, though [26]). The current states of concepts can be characterized by their activity or state vector, $A_i^{(t)}$, where i is the index of the concept, and t stands for time. Using a simple inference formula, e.g. Eq. (1) and an initial state vector $A^{(0)}$ a simulation can be performed and the states of the system over time can be investigated.

$$A_i^{(t+1)} = f\left(\sum_{j=1}^{M} w_{ji} A_j^{(t)}\right) \tag{1}$$

The simulation with a real-valued FCM may behave three different ways: (i) the system converges relatively quickly to a stable, equilibrium state (fixed-point attractor, FP), (ii) after a certain time step starts to repeat a series of state vectors (limit cycle, LC) or (iii) behaves chaotically [27]. In order to constrain activation values into their allowed interval, the application of a suitable $f : \mathbb{R} \rightarrow I_k$ threshold function is also necessary. In most cases, $I_1 = [0, 1]$ (where $k = 1$) is the preferred interval, which can be achieved by e.g. Eq. (2). Another usual case is when $I_2 = [-1, 1]$ (where $k = 2$).

$$f(x) = \frac{1}{1 + e^{-\lambda x}} \tag{2}$$

1.2 The Motivating Problem: Waste Management System

The authors' goal was to model an Integrated Waste Management System (IWMS) with FCM. According to the opinion of experts and the general consensus of the available literature, six main factors (concepts) were identified. Even a survey was composed and sent to the experts to collect their opinion about the direction and strength of the interconnections among these driving factors. The original FCM model of the investigated IWMS was created on the basis of these two sources and simulations were realized.

Later, time series data were also collected from literature by using text mining techniques. It made possible to use machine learning in order to construct a model based on objective data [18]. Unfortunately, a contradiction was found when the two models were compared thus, this simple model of small size proved to be oversimplified and inaccurate [4].

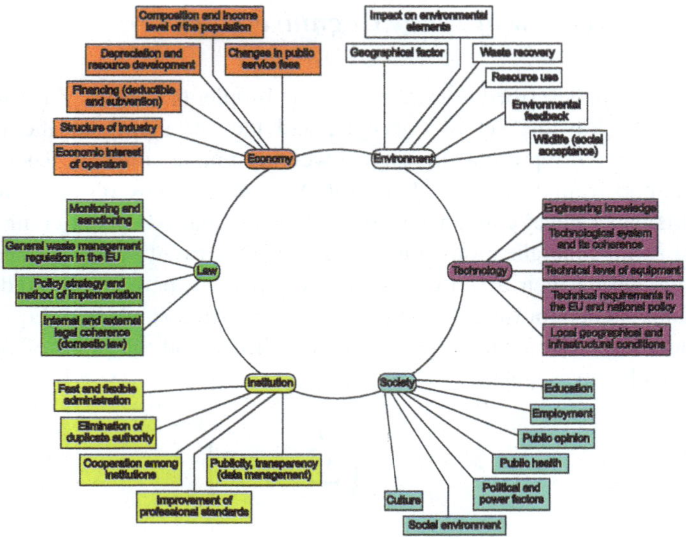

Fig. 1 The six main factors of the original model of IWMS can be seen in the middle of the figure and their subsystems around them

In order to overcome these difficulties, an expert workshop was organized and a more detailed model was elaborated [3]. All factors were split into 4-7 further sub-factors (see Fig. 1, it must be remarked however, that some of the sub-factors showed connection to more than one original concept), and the relationships among new concepts were defined. This model incorporates 33 concepts and 638 arcs, that provides much better accuracy at the cost of very high complexity, therefore it is hardly usable by experts in practice. The need for a reduced model, which provides the same level of accuracy but less complex, was born immediately.

1.3 Existing Model Reduction Methods

Several methods were suggested in the literature to reduce over-sized models. In [2] the model is prepared using historical data and the concepts are clustered with the DEMATEL [5] method. The concepts of the original model are arranged in the plane according to their importance and cause/effect nature. This special arrangement made possible to use two clustering methods as well. One of them uses K-Means clustering and takes only the cause/effect attribute into account. The second method is more complex, applies two stages and uses the importance values, too.

FCMs are used for prediction and knowledge discovery of the HIV-1 drug resistance in [15]. Protease proteins are represented with FCM and causal connections among sequence positions are estimated by Particle Swarm Optimization. The authors look for the strongest sequence positions related to the resistance target with Ant Colony Optimization, and later they remove some concepts and their relationships in order to reduce the complexity of the model while keeping the quality of the map inference at an acceptable level.

The a posteriori removal of weak concepts and connection weights is proposed in [9]. The authors tested the remaining inference capabilities of reduced FCMs with real-world time series.

1.4 Model Reduction Technique Based on Fuzzy Tolerance Relations

The authors already published a model reduction approach in [7] and further analyzed it in [8]. The idea of the method is based on a very essential generalization of the state reduction method for finite state machines and sequential systems with partially specified states [10]. The method creates clusters of concepts using fuzzy tolerance relations (FTR), and uses these clusters as concepts in the reduced model. The cluster members are selected on the basis of their "distance". This distance is based on the strength and direction of causal connections, and can be defined several ways, e.g. with their normalized, squared Euclidean distance (metric "C" in [7]; this distance measure is applied later in this paper).

At first, each concept C_i is wrapped by the cluster K_i. This is called the *initial concept* of the cluster. The cluster is later extended by other concepts not farther then the design parameter $\epsilon \in [0, 1]$ (see function *buildCluster* in [7]). The extreme values of ϵ are obviously useless in practice, because 0 results no reduction at all, and conversely, 1 merges the model into a single cluster, but useful values can be defined only with some expert knowledge, experiences or experimentation.

As a consequence of the properties of FTR, the same concepts may appear in several clusters, moreover, the clusters may not unique, but in the reduced model only one of the equivalent clusters have to be kept. In other respect, the membership of a concept is crisp; it may appear fully in a cluster or not at all.

Finally, the idea of connection weights among clusters have to be defined based on the connections of concepts (function *getWeight* of [7] is applied in this paper, but other definitions are also possible and are already investigated in [8]).

2 Further Clustering-Based Model Reduction Techniques

2.1 K-Means-Based Method

In this paper two clustering-based model reduction methods are presented. One of them is based on K-Means clustering [6], and is thus not entirely new, because this idea has appeared also in [2], only the way of usage is different here. First, data vectors are defined for clustering using the connection weights of concepts. The data vector D_i of concept C_i is a row vector and defined in the following way: $\mathbf{D_i} = [w_{i,*}\ w_{*,i}]$, where $i \in [1, M]$ is the number of the concept, $w_{i,*}$ refers to the outgoing and $w_{*,i}$ to the incoming connection weights of C_i. Using this alternate representation concepts with similar incoming or outgoing weights can be detected by existing clustering methods and wrapped by the same cluster. Self loops are denied in the investigated maps therefore two elements of these vectors are always 0. The proposed method is usable with more general maps as well, however.

Instead of the distance value, ϵ used in the FTR-based reduction method, the K-Means algorithm requires the number of clusters, but both approaches are suitable to control the size of the reduced model. According to the properties of the underlying K-Means algorithm, the concepts appear in exactly one of the clusters, and their membership is crisp. These properties seem to be disadvantageous in contrast to the FTR-based algorithm (which allows the same concept to appear in multiple clusters), but it's easy implementation worth the effort to investigate the results of merger.

2.2 Fuzzy C-Means-Based Method

Against the above KM-based method, the second, Fuzzy C-Means [29] clustering-based (FCM-based) method allows concepts to appear in multiple clusters with different fuzzy membership values, but the sum of these membership values has to be 1. In contrast, the FTR-based method assigns crisp membership values to cluster members (0 or 1), and theoretically the same concept may appear in all clusters.

The data vectors to be clustered can be defined in the very same way, and the number of necessary clusters has to be defined in advance, too. The function to calculate inter-cluster weights is slightly modified, to take the membership values of concepts into account (weighted average is implemented).

$$num = \sum_{i,j,i \neq j} \mu_{a,i}\mu_{b,j}w_{i,j} \tag{3}$$

$$denom = \sum_{i,j,i \neq j} \mu_{a,i}\mu_{b,j} \tag{4}$$

$$w_{a,b} = \begin{cases} num/denom\ if denom \neq 0 \\ 0 \qquad\qquad\quad otherwise \end{cases} \tag{5}$$

In Eqs. (3)–(5), $\mu_{a,i}$ defines the membership value of concept C_i in cluster a, and similarly, $\mu_{b,j}$ defines the membership value of concept C_j in cluster b. $w_{i,j}$ expresses the weight of connection between concepts i and j and $w_{a,b}$ is the weight between clusters a and b.

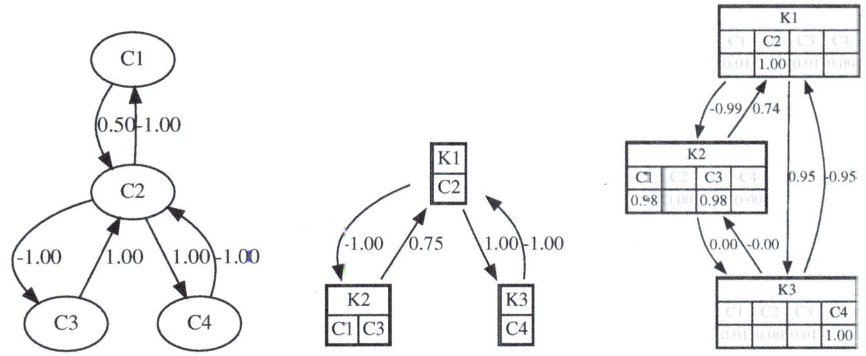

a) The original model to be reduced.

b) Model reduced with the KM-based method.

c) Model reduced with the FCM-based method. All concepts of the original model are merged into the clusters with different membership values.

Fig. 2 Reduction of the basic example model

Table 1 Connection matrix and data vectors of the basic example model

(a) Connection matrix of the basic example model D_i

	C_1	C_2	C_3	C_4
C_1	0	0.5	0	0
C_2	−1	0	−1	1
C_3	0	1	0	0
C_4	0	−1	0	0

(b) data vectors of concepts and their corresponding clusters

Datavector	Cluster								
D_1	K_2	0	0.5	[illegible]	0	0	−1	0	0
D_2	K_1	−1	0	−1	1	0.5	0	1	−1
D_3	K_2	0	1	[illegible]	0	0	−1	0	0
D_4	K_3	0	−1	[illegible]	0	0	1	0	0

Both methods are implemented in Octave, the source codes and data files are available on the Internet [16]. The underlying clustering methods use random numbers, thus the results of clustering may be different, but it does not influence the usability of their result (e.g. the cluster numbers can be any of the possible permutations).

3 Results

3.1 A Basic Example of Model Reduction

In order to illustrate the operation and results of KM-based and FCM-based clustering methods, a very simple model containing four concepts is provided and reduced to 3 clusters (see Fig. 2 and Table 1). Not only the conterts of clusters incluc-ing membership values are included in the figure, but also the connection weights

Table 2 Reduced models containing 18 clusters

Cluster	KM-based	FCM-based	FTR-based
K1	C1.1+C6.5	C1.1+C1.3+C4.7+C6.5	C1.1+C1.2+C1.3+C2.3+C4.3+C4.4+C4.7+C5.1+C6.3
K2	C1.2+C4.3+C6.2+C6.3+C6.4	C1.2+C3.2	C1.1+C1.2+C1.4+C1.5+C3.3+C4.4+C5.2+C5.3+C5.4+C6.4+C6.5
K3	C1.3	C1.4+C1.5	C1.1+C1.2+C1.5+C3.3+C3.4+C3.5+C4.3+C5.4+C6.4
K4	C1.4	C2.1+C2.3	C1.1+C1.2+C3.2+C3.3+C3.4+C3.5+C6.1
K5	C1.5+C4.4+C4.5+C4.6	C2.2	C1.1+C1.3+C1.4+C2.5+C3.1+C3.3+C4.3+C4.4+C4.5+C4.6+C6.1
K6	C2.1	C2.4	C1.1+C1.3+C1.4+C2.5+C3.1+C3.3+C4.3+C4.4+C4.6+C6.1+C6.4
K7	C2.2	C2.5	C1.1+C1.3+C1.4+C3.1+C3.3+C4.3+C6.1+C6.2+C6.3+C6.4
K8	C2.3	C2.6	C1.1+C1.4+C1.5+C2.5+C3.3+C3.5+C4.4+C5.2+C5.3+C6.4
K9	C2.4	C3.1	C1.1+C1.4+C1.5+C4.1+C4.4+C4.5+C4.6+C5.2+C5.3
K10	C2.5+C2.6	C3.3+C3.5+C5.4	C1.1+C1.4+C2.5+C3.1+C3.3+C3.5+C4.3+C4.4+C6.1+C6.4
K11	C3.1+C3.2+C3.4	C3.4	C1.1+C1.4+C2.5+C4.2+C4.4+C4.5+C4.6+C5.2+C5.3
K12	C3.3+C3.5+C3.6+C5.4	C3.6	C1.1+C2.3+C2.5+C3.3+C3.5+C4.3+C4.4+C5.3+C6.4
K13	C4.1	C4.1+C4.4	C1.1+C3.1+C3.2+C3.3+C3.4+C3.5+C6.1
K14	C4.2	C4.2+C5.2+C5.3	C1.1+C3.1+C3.3+C3.4+C3.5+C4.3+C6.1+C6.2+C6.4
K15	C4.7	C4.3+C6.2+C6.3+C6.4	C1.2+C3.3+C3.4+C3.5+C3.6+C5.4
K16	C4.7	C4.5+C5.1	C2.1+C2.3+C2.5+C4.4+C5.1
K17	C5.2+C5.3	C4.6	C2.2+C2.3
K18	C6.1	C6.1	C2.4+C2.6

Table 3 Reduced models containing 23 clusters

Cluster	KM-based	FCM-based	FTR-based
K1	C1.1	C1.1	C1.1–C1.4
K2	C1.2	C1.2+C6.2	C1.1–C1.5
K3	C1.3+C4.7	C1.3	C1.1–C6.5
K4	C1.4	C1.4+C1.5+C4.1+C4.4	C1.2–C6.4
K5	C1.5	C2.1+C2.3	C1.3–C4.7
K6	C2.1	C2.2	C2.1
K7	C2.2	C2.4	C2.2
K8	C2.3	C2.5	C2.3
K9	C2.4	C2.6	C2.4
K10	C2.5	C3.1	C2.5
K11	C2.6	C3.2	C2.6
K12	C3.1	C3.3+C3.4+C3.5	C3.1
K13	C3.2+C3.3+C3.5+C5.4	C3.6	C3.2
K14	C3.4	C4.2+C5.3+C5.4	C3.3–C3.4+C3.5
K15	C3.6	C4.3+C6.3	C3.6
K16	C4.1+C4.2+C4.4+C4.5+C4.6	C4.5	C4.1
K17	C4.3+C6.2+C6.3	C4.6	C4.2–C4.4
K18	C5.1	C4.7	C4.3–C6.1+C6.2+C6.3+C6.4
K19	C5.2	C5.1	C4.4–C4.5+C4.6
K20	C5.3	C5.2	C5.1
K21	C6.1	C6.1	C5.2
K22	C6.4	C6.4	C5.3
K23	C6.5	C6.5	C5.4

Table 4 Reduced models containing 28 clusters

Cluster	KM-based	FCM-based	FTR-based	Cluster	KM-based	FCM-based	FTR-based
K1	C1.1+C5.1	C1.1	C1.1	K15	C3.4	C4.2	C3.4
K2	C1.2	C1.2+C1.5	C1.2	K16	C3.5	C4.3	C3.6
K3	C1.3+C4.7	C1.3	C1.3	K17	C3.6	C4.4	C4.1
K4	C1.4	C1.4	C1.4	K18	C4.1	C4.5	C4.2
K5	C1.5	C2.1+C2.3	C1.5	K19	C4.2	C4.6	C4.3+C6.2+C6.3+C6.4
K6	C2.1	C2.2	C2.1	K20	C4.3+C6.1	C4.7+C5.3	C4.4+C4.5
K7	C2.2	C2.4	C2.2	K21	C4.4	C5.1	C4.5+C4.6
K8	C2.3	C2.5	C2.3	K22	C4.5	C5.2	C4.7
K9	C2.4	C2.6	C2.4	K23	C4.6	C5.3	C5.1
K10	C2.5	C3.1	C2.5	K24	C5.2	C5.4	C5.2
K11	C2.6	C3.2	C2.6	K25	C5.3	C6.1	C5.3
K12	C3.1	C3.3+C3.4	C3.1	K26	C6.2	C6.2	C5.4
K13	C3.2	C3.5+C3.6	C3.2	K27	C6.3+C6.4	C6.4	C6.1
K14	C3.3+C5.4	C4.1	C3.3+C3.5	K28	C6.5	C6.5	C6.5

calculated with the appropriate function (*getWeight*) of the FTR-based method. In Fig. 2c, membership values less than 0.5 are grayed out.

3.2 Reduction of the Waste Management Model

In order to check the correct operation of the proposed new reduction methods and to compare their reduced models with the ones created by the FTR-based method, a much complex model, the model of our motivating problem was also reduced. The model originally contained 33 concepts was reduced to 18 ($\epsilon = 0.048$), 23 ($\epsilon = 0.024$) and 28 ($\epsilon = 0.018$) clusters, and the clusters of the reduced models are detailed in Tables 2, 3 and 4, respectively. In the case of FCM-based reduction, due to space limitations, the exact membership values are not provided, but only concepts with greater than 0.5 membership values are indicated as cluster members.

According to our experiences, membership values are very close to 0 or 1 in most cases (e.g. concepts with high membership values are greater than or equal to 0.994 in the investigated cases). Another interesting issue is that the default parameter value of the objective function of the Fuzzy C-Means algorithm ($m = 2$) resulted in an unusable reduced model, where the membership values were all the same. It was certainly the effect of long, 66 element data vectors. The value of $m = 1.1$ seemed to lead to acceptable results, and slight modifications of it did not modify the results significantly. FCM-based clusters indicated in the tables were also created using the parameter value $m = 1.1$.

The clusters created with the applied three methods are rather different. The opinion of an expert group could be asked about the clarity and usefulness of these model versions, but it would be much better to perform an exhaustive behavioral analysis and comparison in the near future.

4 Conclusions

Two novel, clustering-based Fuzzy Cognitive Map model reduction methods were proposed in this paper. According to our hypothesis based on the properties of K-Means clustering, the first new model reduction technique provides reduced models with less accuracy than the other, Fuzzy C-Means based method or the already proposed method based on fuzzy tolerance relations. The FCM-based reduction method has the potential to represent the real importance of concepts in a cluster better, however. In order to confirm our hypothesis additional investigations are required, e.g. the behavior of original and reduced models should be compared using high number of initial state vectors, and the results of simulations should be thoroughly compared. The effect of parameter m of Fuzzy C-Means clustering on model reduction should be also investigated. That is going to be our next task.

The combination of the FTR-based and the FCM-based methods could be potentially even better, however: the membership value of a specific concept should be allowed to be in the [0, 1] interval in each cluster. The extension of FTR-based method seems also possible based on the already calculated distance and ϵ values.

References

1. Ahmadi, S., Papageorgiou, E.I., Yeh, C.H., Martin, R.: Managing readiness-relevant activities for ERP implementation. Comput. Ind. https://doi.org/10.1016/j.compind.2014.12.009 (in press) (2014)
2. Alizadeh, S., Ghazanfari, M., Fathian, M.: Using data mining for learning and clustering FCM. Int. J. Comput. Intell. **4**(2), 118–125 (2008)
3. Buruzs, A., Hatwágner, M., Torma, A., Kóczy, L.: Expert based system design for integrated waste management. Int. Sch. Sci. Res. Innov. **8**(12), 685–693 (2014)
4. Buruzs, A., Hatwágner, M.F., Kóczy, L.T.: Fuzzy cognitive maps and bacterial evolutionary algorithm approach to integrated waste management systems. J. Adv. Comput. Intell. Intell. Inform. **18**(4), 538–548 (2014)
5. Gabus, A., Fontela, E.: Perceptions of the world problematique: communication procedure, communicating with those bearing collective responsibility (dematel report no.1). Technical Report, Battelle Geneva Research Centre, Geneva, Switzerland (1973)
6. Hartigan, J.A., Wong, M.A.: Algorithm as 136: a K-means clustering algorithm. J. R. Stat. Society. Ser. C (Appl. Stat.) **28**(1), 100–108 (1979)
7. Hatwagner, M.F., Koczy, L.T.: Parameterization and concept optimization of FCM models. In: 2015 IEEE International Conference on Fuzzy Systems (FUZZ-IEEE), pp. 1–8. IEEE, Istanbul (2015)
8. Hatwágner, M.F., Yesil, E., Dodurka, M.F., Papageorgiou, E., Urbas, L., Kóczy, L.T.: Two-stage learning based fuzzy cognitive maps reduction approach. IEEE Trans. Fuzzy Syst. **26**(5), 2938–2952 (2018)
9. Homenda, W., Jastrzebska, A., Pedrycz, W.: Computer Information Systems and Industrial Management: 13th IFIP TC8 International Conference, CISIM 2014, Ho Chi Minh City, Vietnam, November 5–7, 2014. Proceedings, chap. Time Series Modeling with Fuzzy Cognitive Maps: Simplification Strategies, pp. 409–420. Springer, Berlin, Heidelberg (2014)
10. Kohavi, Z., Jha, N.K.: Switching and Finite Automata Theory, 3rd edn. Cambridge University Press (2009)
11. Kontogianni, A.D., Papageorgiou, E.I., Tourkolias, C.: How do you perceive environmental change? fuzzy cognitive mapping informing stakeholder analysis for environmental policy making and non-market valuation. Appl. Soft Comput. **12**(12), 3725–3735 (2012)
12. Kosko, B.: Fuzzy cognitive maps. Int. J. Man-Mach. Stud. **24**, 65–75 (1986)
13. Kosko, B.: Neural Networks and Fuzzy Systems. Prentice-Hall, New Jersey (1992)
14. Kosko, B.: Hidden patterns in combined and adaptive knowledge networks. Int. J. Approx. Reason. **2**(4), 377–393 (1988)
15. Nápoles, G., Grau, I., Bello, R., Grau, R.: Two-steps learning of fuzzy cognitive maps for prediction and knowledge discovery on the HIV-1 drug resistance. Expert Syst. Appl. **41**(3), 821–830 (2014). https://doi.org/10.1016/j.eswa.2013.08.012, Methods and Applications of Artificial and Computational Intelligence
16. Octave scripts and data files for conference paper presented at escim2019. http://rs1.sze.hu/~hatwagnf/escim2019/escim2019.zip, accessed: 2019-05-31
17. Papageorgiou, E.I.: A new methodology for decisions in medical informatics using fuzzy cognitive maps based on fuzzy rule-extraction techniques. Appl. Soft Comput. **11**, 500–513 (2011)

18. Papageorgiou, E.I.: Learning algorithms for fuzzy cognitive maps—a review study. IEEE Trans. Syst. Man Cybern. Part C (Appl. Rev.) **42**(2), 150–163 (2012)
19. Papageorgiou, E.I.: Fuzzy Cognitive Maps for Applied Sciences and Engineering: From Fundamentals to Extensions and Learning Algorithms, vol. 54. Springer Science & Business Media (2013)
20. Papageorgiou, E.I.: Review study on fuzzy cognitive maps and their applications during the last decade. In: Business Process Management, pp. 281–298. Springer (2013)
21. Papageorgiou, E.I., Salmeron, J.L.: Methods and algorithms for fuzzy cognitive map-based decision support. In: Papageorgiou, E.I. (ed.) Fuzzy Cognitive Maps for Applied Sciences and Engineering (2013)
22. Papageorgiou, E.I., Hatwágner, M.F., Buruzs, A., Kóczy, L.T.: A concept reduction approach for fuzzy cognitive map models in decision making and management. Neurocomputing **232**, 16–33 (2017)
23. Pedrycz, W.: The design of cognitive maps: a study in synergy of granular computing and evolutionary optimization. Expert Syst. Appl. **37**, 7288–7294 (2010)
24. Salmeron, J.L.: Supporting decision makers with fuzzy cognitive maps. Res.-Technol. Manag. **52**(3), 53–59 (2009)
25. Stach, W., Kurgan, L., Pedrycz, W.: A survey of fuzzy cognitive map learning methods. Issues in soft computing: theory and applications, pp. 71–84 (2005)
26. Stylios, C.D., Groumpos, P.P., et al.: Mathematical formulation of fuzzy cognitive maps. In: Proceedings of the 7th Mediterranean Conference on Control and Automation, pp. 2251–2261 (1999)
27. Tsadiras, A.K.: Comparing the inference capabilities of binary, trivalent and sigmoid fuzzy cognitive maps. Inf. Sci. **178**(20), 3880–3894 (2008)
28. Vliet, M.v., Kok, K., Veldkamp, T.: Linking stakeholders and modellers in scenario studies: the use of fuzzy cognitive maps as a communication and learning tool. Futures **42**(1), 1–14 (2010)
29. Yen, J., Langari, R.: Fuzzy Logic: Intelligence, Control, and Information. Prentice Hall (1999). https://books.google.hu/books?id=XnRQAAAAMAAJ
30. Zhang, W.R.: Bipolar fuzzy sets. In: The 1998 IEEE International Conference on Fuzzy Systems Proceedings. IEEE World Congress on Computational Intelligence, vol. 1, pp. 835–840. IEEE (1998)

Some Implications of Interval Approach to Dimension for Network Complexity

Věra Kůrková

Abstract An interval approach to the concept of dimension is presented. Implications of exponentially growing quasiorthogonal dimension for estimates of network complexity are analyzed. Bounds on correlations of computational tasks represented by high-dimensional vectors are derived. Network complexity is explored from the point of view of the concentration of measure phenomenon.

Keywords Quasiorthogonal dimension · Sparsity of feedforward networks · High-dimensional geometry · Concentration of measure · Covering numbers

1 Introduction

The intuitive concept of dimension have been formalized mathematically in many ways. A topological approach, based on the recursive character of d-dimensional objects having $d-1$-dimensional boundaries, was proposed by Poincaré. A metric version of dimension was developed by Hausdorff and popularized by Mandelbrot as fractal dimension, which can even have nonintegral values. The most simple among these formalizations is the geometric concept of dimension defined as the maximal number of orthogonal vectors, i.e., vectors with angular distances $\pi/2$. The orthogonal dimension of a Euclidean space is equal to the maximal number of linearly independent vectors, which is an algebraic concept defining the dimension of a vector space.

Replacing the exact orthogonality with a condition that the angular distances between vectors are within a fixed closed symmetric interval around $\pi/2$, we defined in [1] the concept of quasiorthogonal dimension. We showed that relaxing the crisp condition of exact orthogonality with a fuzzy one of ε-quasiorthogonality considerably increases the number of highly uncorrelated vectors. We proved in [1] that for

V. Kůrková (✉)
Institute of Computer Science of the Czech Academy of Sciences, Pod Vodárenskou věží 2, 18207, Prague, Czech Republic
e-mail: vera@cs.cas.cz

M. E. Cornejo et al. (eds.), *Computational Intelligence and Mathematics for Tackling Complex Problems 2*, Studies in Computational Intelligence 955,
https://doi.org/10.1007/978-3-030-88817-6_13

a fixed size of an interval $[-\varepsilon, \varepsilon]$, the ε-quasiorthogonal dimension grows exponentially with an increasing orthogonal dimension.

Many classification and regression tasks can be modeled by functions on finite domains, which are typically quite large. They can be formed by pixels of pictures, discretized cubes, or scattered vectors of data. Functions on finite domains can be represented as vectors in Euclidean spaces of dimensions equal to cardinalities of their domains. The exponential growth of quasiorthogonal dimension implies that there are large numbers of highly uncorrelated functions even on domains of moderate sizes.

Feedforward neural networks compute parameterized families of input-output functions. It has long been known that under a mild condition on a dictionary of computational units, feedforward networks have the universal representation property, i.e., they can exactly compute any real-valued function on a finite domain [2]. However, the arguments proving this property assume that the number of network units is potentially as large as the size of the domain. For large domains, such networks might be too large for efficient implementations. Thus it is desirable to investigate classes of functions which can be computed by reasonably sparse networks.

In this paper, we investigate how exponential growth of the number of highly uncorrelated tasks imposes limitations on capabilities of feedforward networks to compute classes of regression and classification tasks. We show that critical factors are covering numbers of dictionaries of computational units and a prior knowledge of computational tasks.

The paper is organized as follows. In Sect. 2, we introduce the interval approach to dimension and discuss its relationship to geometry of high-dimensional Euclidean spaces. Section 3 is devoted to complexity issues of feedforward networks. In Sect. 4, some estimates of approximate measures of network sparsity are derived using properties of high-dimensional spaces. Section 5 is a brief discussion.

2 Quasiorthogonal Dimension

Let $\mathbb{R}^m$ denote the m-dimensional Euclidean space, S^{m-1} the unit sphere in $\mathbb{R}^m$, and $\langle u, v \rangle$ the inner product of $u, v \in \mathbb{R}^m$. For $\varepsilon > 0$ and the corresponding angle $\alpha = \arccos \varepsilon$, a set T of vectors in $\mathbb{R}^m$ is called ε-quasiorthogonal if the scalar product of each pair of distinct normalized vectors in T is smaller than ε. While there are only m exactly orthogonal unit vectors in the m-dimensional Euclidean space, the number of ε-quasiorthogonal vectors grows with m exponentially. In [1] we proved the lower bound $e^{\frac{m\varepsilon^2}{2}}$ on the *quasiorthogonal dimension* $\dim_\varepsilon m$ defined as the maximal number of ε-quasiorthogonal vectors. Using the concept of the clique number from graph theory, we proved even a stronger result stating that this lower bound holds even for ε-quasiorthogonal vectors with entries 1 and -1 (which are elements of the Hamming cube). Toleration of a small inaccuracy within an interval $[-\varepsilon, \varepsilon]$ in scalar products largely increases the number of highly uncorrelated vectors.

The exponential growth of the quasiorthogonal dimension can be explained by properties of high-dimensional Euclidean spaces. With increasing dimension m, relative areas occupied on spheres S^{m-1} by "polar caps" decrease exponentially fast to zero. More precisely, let

$$C(v, \varepsilon) = \{u \in S^{m-1} \mid \langle u, v \rangle \geq \varepsilon\},$$

denotes the *polar cap* centered at a fixed unit vector v, which contains all unit vectors u which have the angular distance from v at most $\alpha = \arccos \varepsilon$ (the inner product $\langle u, v \rangle$ is at least ε). It can be shown using integration in spherical polar coordinates that for a fixed $\varepsilon > 0$, the normalized Lebesgue measure μ of the polar cap $C(v, \varepsilon)$ decreases exponentially fast to zero with the dimension m increasing as

$$\mu(C(v, \varepsilon)) \leq e^{-\frac{m\varepsilon^2}{2}} \tag{1}$$

(see, e.g., [3]). So all vectors in the complement of the union of the two polar caps $C(v, \varepsilon) \cup C(-v, \varepsilon)$ are ε-quasiorthogonal to v. The upper bound $e^{-\frac{m\varepsilon^2}{2}}$ on the relative measure of the polar cap $C(v, \varepsilon)$ implies that one can pack exponentially many disjoint caps in the sphere S^{m-1} in the m-dimensional space $\mathbb{R}^m$. For ε corresponding to the angle $\alpha = \arccos \varepsilon$, this packing number gives the lower bound on quasiorthogonal dimension.

A generalization of the upper bound (1) obtained by replacing the inner product $\langle ., v \rangle$ with a general Lipschitz function gives the Lèvy Lemma (see, e.g., [4]). It states that almost all values of a Lipschitz function on a high-dimensional sphere are close to their median. This property of high-dimensional spheres is called the *concentration of measure phenomenon*. Similar property was also discovered in probability theory, where it has been studied in terms of bounds on large deviations of sums or even certain sufficiently smooth functions of several random variables see, e.g., Hoeffding [5], Chernoff [6], and Azuma [7]. It implies that randomized techniques work almost deterministically [8]. Concentration of measure is also the essence of the proof of the Johnson-Lindenstrauss Flattening Lemma. It guarantees a possibility of dimension reduction of d-dimensional data by a random projection to a lower dimension bounded from below by $\frac{8}{\varepsilon} \log d$ such that the projection is a near-isometry (preserves distances within a multiplicative factor $1 \pm \varepsilon$) (see, e.g., [4]).

3 Computation of Finite Mappings by Feedforward Networks

Feedforward networks are computational models that compute parameterized families of functions. Domains of functions modeling some practical classification or regression tasks are finite subsets of $\mathbb{R}^d$. Thus input-output functions of feedforward networks can be seen as vectors in $\mathbb{R}^m$, where m is the cardinality of the function domain X. Often, X is quite large. For $X \subset \mathbb{R}^d$ we denote by

$$\mathcal{F}(X) := \{f : X \to \mathbb{R}\}$$

the set of all real-valued functions on X and by

$$S_1(X) := \{f \in \mathcal{F}(X) \mid \|f\|_2 = 1\}$$

the unit sphere in $\mathcal{F}(X)$.

We focus on *feedforward networks with single linear outputs*. Such networks compute input-output functions from the set

$$\operatorname{span} G := \left\{ \sum_{i=1}^{n} w_i g_i \,\middle|\, w_i \in \mathbb{R},\ g_i \in G,\ n \in \mathbb{N} \right\} \tag{2}$$

where G, called a *dictionary*, is a parameterized family of functions. In *shallow (one-hidden-layer) networks*, G is formed by functions computable by a given type of computational units, whereas in *deep networks* with several hidden layers, it is formed by combinations and compositions of functions representing units from lower layers.

It has long been known (see, e..g., [2]) that many feedforward networks have the *universal representation property*, i.e., they can exactly compute any function on a finite domain. On finite domains, typically, dictionaries are linearly dependent (they are called *overcomplete*). For a function f on X and a dictionary G, we denote

$$W_f(G) := \{w = (w_1, \ldots, w_n) \in \mathbb{R}^n \mid f = \sum_{i=1}^{n} w_i g_i, g_i \in G, n \in \mathbb{N}\}$$

the set of output-weight vectors of networks with units from G representing f. For classes of networks with the universal representation capability, the sets $W_f(G)$ are non empty for all $f \in \mathcal{F}(X)$.

It is desirable to find among all representations of f as an input-output function of a network with units from G the most sparse ones, i.e., to find in the set $W_f(G)$ vectors with the minimal number of non-zero entries. Formally, for a vector $w \in \mathbb{R}^n$, the *number of its non-zero entries* is denoted $\|w\|_0$. It is called "*l_0-pseudonorm*", but it is neither a norm nor a pseudonorm. Minimization of "l_0-pseudonorm" is a difficult non convex problem which has been studied in signal processing and was proven to be NP-hard in some cases [9].

In neurocomputing, instead of "l_0-pseudonorm", l_1 and l_2-norms have been used as stabilizers in weight-decay regularization methods [10]. Acting as a stabilizer, l_2-norm penalizes even a small number of large output weights but it can tolerate many small ones. On the other hand, l_1-norm stabilizers penalize many small output weights as well as few large ones. In addition to approximating well the convexification of

l_0 and penalizing many small output weights, l_1-norm have several other properties which make it a good approximate measure of sparsity. The minimal value of the l_1-norm of an output-weight vector of a network computing a function f plays a role of an upper bound on rates of approximation by networks with increasing "l_0-pseudonorms" of output-weight vectors. It can be expressed in terms of a norm generated by a dictionary G called *G-variation*. It is defined for a bounded subset G of a normed linear space $(\mathcal{X}, \|.\|)$ as

$$\|f\|_G := \inf\left\{c \in \mathbb{R}_+ \;\middle|\; f/c \in \mathrm{cl}_{\mathcal{X}} \,\mathrm{conv}\,(G \cup -G)\right\},$$

where $-G := \{-g \mid g \in G\}$, $\mathrm{cl}_{\mathcal{X}}$ denotes the closure with respect to the topology induced by the norm $\|\cdot\|_{\mathcal{X}}$, and conv is the convex hull. If the set over which the infimum is taken is empty, then $\|f\|_G := \infty$. Variation with respect to Heaviside perceptrons (called *variation with respect to half-spaces*) was introduced in [11] and extended to general dictionaries in [12] (for more properties of G-variation see [13]). Denoting

$$W_f(G)_1^* := \{w^* \in W_f(G) \mid \|w^*\|_1 = \min_{w \in W_f(G)} \|w\|_1\}$$

we obtain the following proposition.

Proposition 1 *Let $X \subset \mathbb{R}^d$, G be a bounded subset of $\mathcal{F}(X)$, and $f \in \mathcal{F}(X)$ be such that $\|f\|_G$ is finite. Then $\|f\|_G \leq \|w^*\|_1$ for all $w \in W_f(G)_1^*$.*
When G is finite, then $\|f\|_G = \|w^*\|_1$ for all $w \in W_f(G)_1^*$.

Thus the minimal l_1-norm of output-weight vectors of a representation of a function by a network is a good *approximate measure of network sparsity*. The next geometric characterization of variational norm from [13] provides a tool for obtaining lower bounds on this approximate measure. By $G^\perp$ is denoted the *orthogonal complement of* G in the Hilbert space $\mathcal{F}(X)$.

Theorem 1 *Let X be a finite subset of $\mathbb{R}^d$ and G be a bounded subset of $\mathcal{F}(X)$. Then for every $f \in \mathcal{F}(X) \setminus G^\perp$, $\|f\|_G \geq \frac{\|f\|^2}{\sup_{g \in G} |\langle g, f\rangle|}$.*

Theorem 1 gives an insight into the concept of variational norm. It implies that functions which are "nearly orthogonal" to all elements of a dictionary have large variations and thus cannot be computed by networks having "small" l_1-norms of output-weigh vectors.

4 Bounds on Approximate Measure of Network Sparsity

Combining the geometric characterization of variational norm from Theorem 1 with small sizes of polar caps guaranteed by the upper bound (1), we obtain an estimate of the relative area of the unit sphere $S_1(X)$ formed by functions for which all representations by networks with units from a dictionary G computing must satisfy a lower bound on l_1-norms of output-weight vectors. The following bound is expressed in terms of a size of a dictionary measured by its covering numbers.

For $\varepsilon > 0$, an *ε-net in F* is a set $\{f_1, \ldots, f_n\} \subseteq \mathcal{X}$ (where $\mathcal{X}$ is a metric space), such that the family of the closed balls $B_\varepsilon(f_i)$ of radii ε centered at f_i covers F. The *ε-covering number* denoted $\mathcal{N}_\varepsilon(F)$ of a subset F of a metric space $\mathcal{X}$ is the cardinality of a minimal ε-net in F, i.e.,

$$\mathcal{N}_\varepsilon(F) = \min\Big\{n \in \mathbb{N}_+ \mid F \subseteq \bigcup_{i=1}^{n} B_\varepsilon(f_i),\ (\forall i = 1, \ldots, n)(f_i \in F)\Big\}.$$

When the set over which the minimum is taken is empty, then $\mathcal{N}_\varepsilon(F) = +\infty$.

Theorem 2 *Let d be a positive integer, $X \subset \mathbb{R}^d$ with* card $X = m$, *$b > 0$, and $G \subset S_1(X)$ has finite covering numbers. Then*

$$\mu\big[f \in S_1(X) \mid (\forall w \in W_f(G))\ \|w\|_1 \geq b\big] \geq 1 - 2\mathcal{N}_{\arccos(2/b)}(G)\, e^{-\frac{2m}{b^2}}.$$

Because the sizes of polar caps on high-dimensional spheres are small (decrease exponentially fast with m), to contain input-output functions approximating well all functions in $S_1(X)$, a class of networks must have large covering numbers (growing exponentially with m). Note that for some values of ε, covering numbers of finite sets can even be smaller than their sizes. This can happen for highly coherent sets. Theorem 2 implies that for dictionaries having covering numbers growing with sizes of domains only polynomially, most functions are nearly orthogonal to all their elements and their representations require large numbers of units or some output weight must be large (which leads to instability). Both are not desirable.

Thus estimates of covering numbers of various dictionaries used in neurocomputing are important. Some estimates of covering numbers of dictionaries are known for shallow networks, where dictionaries are formed by functions computable by basic computational units. More complicated dictionaries formed by compositions of functions which are used in deep networks are much less understood. Examples of dictionaries formed by binary valued functions are dictionaries of Heaviside or signum perceptrons. Estimates of their sizes follow from the upper bound $2\frac{m^d}{d!}$ on the number of linearly separable dichotomies of m points in $\mathbb{R}^d$ [14]. Thus its size grows with m only polynomially with the polynomial degree equal to d.

5 Discussion

Theorem 2 applies to situations when we have no prior knowledge on the task and thus we have to search for a class of networks potentially capable of processing any uniformly randomly chosen function. However, in real applications, relevance of most functions for a choice of a class of networks suitable for an efficient computation is very low or negligible. This explains many successful applications of shallow neural networks with relatively small dictionaries such as perceptrons or kernel units used in SVM. In [15] we introduced a probabilistic approach modeling a prior knowledge that a binary classification is likely to occur in a given application. We derived some estimates for product probabilities. More general probability distributions and regression tasks are subject of our future research.

Acknowledgements V.K. was partially supported by the Czech Grant Foundation grant GA18-23827S and institutional support of the Institute of Computer Science RVO 67985807.

References

1. Kainen, P.C., Kůrková, V.: Quasiorthogonal dimension of Euclidean spaces. Appl. Math. Lett. **6**, 7–10 (1993)
2. Ito, Y.: Finite mapping by neural networks and truth functions. Math. Sci. **17**, 69–77 (1992)
3. Ball, K.: An elementary introduction to modern convex geometry. In: Levy, S. (ed.) Flavors of Geometry, pp. 1–58. Cambridge University Press (1997)
4. Matoušek, J.: Lectures on Discrete Geometry. Springer, New York (2002)
5. Hoeffding, W.: Probability inequalities for sums of bounded random variables. J. Am. Stat. Assoc. **58**, 13–30 (1963)
6. Chernoff, H.: A measure of asymptotic efficiency for tests of a hypothesis based on the sum of observations. Ann. Math. Stat. **23**, 493–507 (1952)
7. Azuma, K.: Weighted sums of certain dependent random variables. Tohoku Math. J. **19**, 357–367 (1967)
8. Dubhashi, D., Panconesi, A.: Concentration of Measure for the Analysis of Randomized Algorithms. Cambridge University Press (2009)
9. Tillmann, A.: On the computational intractability of exact and approximate dictionary learning. IEEE Signal Process. Lett. **22**, 45–49 (2015)
10. Fine, T.L.: Feedforward Neural Network Methodology. Springer, Berlin (1999)
11. Barron, A.R.: Neural net approximation. In: Narendra, K.S. (ed.) Proceedings of the 7th Yale Workshop on Adaptive and Learning Systems, pp. 69–72. Yale University Press (1992)
12. Kůrková, V.: Dimension-independent rates of approximation by neural networks. In: Warwick, K., Kárný, M. (eds.) Computer-Intensive Methods in Control and Signal Processing. The Curse of Dimensionality, pp. 261–270. Birkhäuser, Boston (1997)
13. Kůrková, V.: Complexity estimates based on integral transforms induced by computational units. Neural Netw. **33**, 160–167 (2012)
14. Schläfli, L.: Gesamelte Mathematische Abhandlungen, Band 1. Birkhäuser, Basel (1950)
15. Kůrková, V., Sanguineti, M.: Classification by sparse neural networks. IEEE Trans. Neural Netw. Learn. Syst. **30**(9), 2746–2754 (2019)

Social Indexes Segregation Based on MEOWA and MOOWA Aggregation Operators

Nuria Martinez, Daniel Gomez, Karina Rojas, Pablo Olaso, and Javier Montero

Abstract Indexes are very used in social sciences. The construction of an index involves aggregating information about variables that sometimes cannot be ranked, being OWA operators the most appropriate alternative. The associated weights of OWA operators can be defined in different ways. In this paper we propose to use MOOWA aggregation operators (maximizing ordinal dispersion instead of entropy) for the construction of indexes where no ranking amongst their variables is possible.

Keywords OWA operators · Ordinal dispersion · Social indexes

1 Introduction

One of the main objectives of social sciences is associated with the construction of indexes for a better understanding of a specific problem. However, most of the indices that have been constructed for this purpose do not admit vagueness when the information is modelled. Classical indexes are based on clear techniques that are based on linear aggregation models in which the variables are weighted according to their importance. In an index (or subindex) construction process, the structure of

Supported by DATA SCIENCE AND SOFT COMPUTING FOR SOCIAL ANALYTIC AND DECISION AID Research Group and by means of grant PGC2018-096509-B-I00 of the Government of Spain

N. Martinez · J. Montero
Facultad de Ciencias Matematicas, Universidad Complutense de Madrid, Madrid, Spain
e-mail: monty@mat.ucm.es

D. Gomez (✉)
Facultad de Estudios Estadisticos, Universidad Complutense de Madrid, Madrid, Spain
e-mail: dagomez@estad.ucm.es

K. Rojas · P. Olaso
Escuela Tecnico Superior de Ingeniería Informática, Seville, Spain
e-mail: karinah.rpatuelli@urjc.es

M. E. Cornejo et al. (eds.), *Computational Intelligence and Mathematics for Tackling Complex Problems 2*, Studies in Computational Intelligence 955,
https://doi.org/10.1007/978-3-030-88817-6_14

the input information (variables) are not necessary ordered as a linear order and thus not every weighted aggregation process will be appropriate. When an index is built based on a weighted linear aggregation model the variables are associated with a weighted so it is clear that the variables are ranked (as a linear order) from the most important variable (in the index construction) to the least important one. And as it is pointed out in [1] (in which a hierarchical social index is built based on fuzzy sets), the variables associated with the subindexes of each hierarchical level cannot be ordered since all of them have the same importance.

Taking all this into account in this in this paper we propose the use of OWA aggregation operators [2] for the construction of these indexes or subindexes in which it is not possible to rank the variables since all of them present the same importance. A key application problem for the use of OWA operators is the determination of the associated weights.

This paper is organized as follows. In the next section we recall the definition of Ordered Weighted Averaging (OWA) aggregation operators and the problem of determining the associated weights. In Sect. 3 we present the index building method as an aggregation process. We will begin with the characterization of a prioritized hierarchical information, and then define a typology of indicators based on their weighting scheme. Finally we present a real case in which a subindex in built based on this methodology. We will conclude the paper with some final remark.

2 OWA and MEOWA Aggregation Operators

An aggregation operator is usually defined (see [1, 3–6]) as a real function $A : [0, 1]^n \rightarrow [0, 1]$, such that for n items in [0, 1], yields an aggregation value in the same interval and fulfilling the following conditions:

- if $x_i \leq y$, then $A(x_1, \ldots, x_n) \leq A(x_1, \ldots, x_{i-1}, y, x_{i+1}, \ldots, x_n)$
- $A, 0, \ldots, 0) = 0$ and $A(1, \ldots, 1) = 1$.

An OWA aggregation operator of dimension n is a special class of aggregation operators that can be defined as a mapping $F : [0, 1]^n \longrightarrow [0, 1]$ with an associated weight vector $w = (w_1, \ldots, w_n)$ having the properties

$$\sum_{i=1}^{n} w_i = 1; \quad 0 \leq w_i \leq 1 \ \forall i = 1, \ldots n.$$

and such that

$$F(x_1, \ldots, x_n) = \sum_{i=1}^{n} w_i x_{(i)}$$

with $x_{(i)}$ being the ith largest component of the x.

In order to determine the weights that represent the importance of the ith largest element the concept of orness is introduced. The Orness measure represents in some way the degree in which an operator is disjunctive. Formally, given an OWA aggregation operator, the orness is defined as

$$orness(W) = \sum_{i=1}^{i=n} \frac{n-i}{n-1} w_i.$$

Compensative connectives have the property that a higher degree of satisfaction of one of the criteria can compensate a lower degree of satisfaction of another criterion. "Or-ing" the criteria means full compensation and "and-ing" the criteria means no compensation.

In [7] the weights are determined by maximizing the entropy of the weight vector for a given level of orness. This family of OWA aggregation operators are known as MEOWA aggregation operators.

3 MOOWA Aggregation Operators

In [8], a different way to obtain the weights of an OWA aggregation operator by maximizing the ordinal dispersion instead of the entropy is introduced. Entropy can be viewed as a dispersion measure of nominal variables and the weights of the OWA as a probability distribution. In this sense, entropy could seem not to be the most adequate measure to capture the dispersion since the weights do not correspond to a nominal variable. To this aim, the MOOWA are defined in a similar way as the MEOWA but maximizing an ordinal dispersion measure.

In this work, we will obtain the MEOWA aggregation operators based on the ordinal dispersion measure defined by Berry and Mielke [9], usually called as IOV and defined as follows.

Given an ordinal variable with values $X = \{1, \ldots, n\}$ and a relative frequency vector $f = (f_1, \ldots, f_n)$, the ordinal dispersion measure IOV is defined as

$$IOV = \sum_{i=1}^{n-1} \sum_{j=i+1}^{n} f_i f_j (j-i). \quad (1)$$

Formally, the mathematical programming problem that we will solve to obtain the MOOWA aggregation operators (for a fixed livel of orness α) can be expressed as follows:

$$\begin{aligned} maximize\ dispersion(w) &= \sum_{i=1}^{i=n-1} \sum_{j=i}^{i=n} (j-i) w_i w_j \\ s.a. \quad & \frac{1}{n-1} \sum_{i=1}^{i=n} (n-i) w_i = \alpha \\ & \sum_{i=1}^{i=n} w_i = 1 \\ & 0 \le w_i \le 1,\ i = 1, \ldots, n. \end{aligned} \tag{2}$$

In [8], it is proven that the vector weight $w = (\alpha, 0, \ldots, 0, 1 - \alpha)$ is the optimal weight vector that maximizes the ordinal dispersion.

4 Social Indexes Segregation Based on MEOWA, MOOWA and OWA Aggregation Operators

A prioritized hierarchical structure of data represents a situation in which there are some clusters of the variables or criteria set $H_1, \ldots H_r$, with respect to which some preferences are clear. The criteria that belong to H_r are more important in the aggregation process than the criteria (or variables) that belong to H_{r-1}, and so on until H_1. Elements or variables of the data associated to each cluster are unstructured since no preferences regarding the aggregation process are given a priori for them.

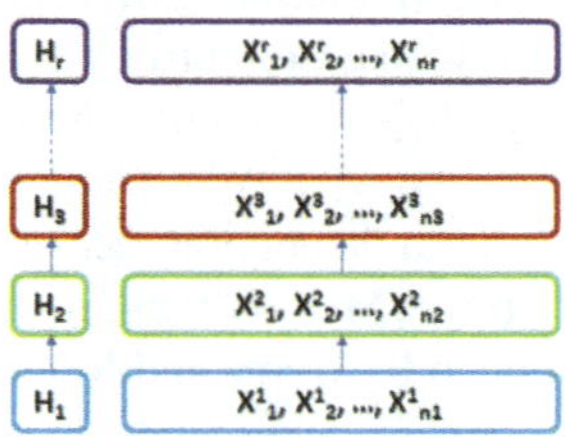

Assuming that we summarize this information in a matrix from $X_{m\times n}$ in such a way x_{ij} represent the valuation of the item i in the variable X_j, a prioritized hierarchical structure of data can be defined as follows:

Definition 1 [1]
Let X be a set of data. X has a *prioritized hierarchical structure* if it is possible to classify the elements of X into r clusters $H_1, \ldots H_r$, satisfying the following conditions:

- $\{H_1, \ldots H_r\}$ constitute a partition of X, i.e. $\bigcup_{i=1}^{r} H_i = X$, and $H_i \cap H_j = \emptyset$, if $i \neq j$.
- A linear order (in fact is also a pre-linear) is defined regarding the subsets $H_1, \ldots H_r$, in the following sense: Given two elements $x, y \in X$, if $x \in H_i$, and

Citizen	I feel represented by some political party	My vote may influence	Political parties take into account the interests of citizens	Candidates comply with their program when elected	I've always voted in the elections.
1	1	4	2	1	4
2	4	5	3	4	5
3	3	3	2	2	2
4	4	2	2	2	4
5	2	2	3	4	3

Fig. 1 Input information

Citizen	Arithmetical mean	MOOWA				MEOWA			
		α=0,3	α=0,5	α=0,7	α=0,9	α=0,3	α=0,5	α=0,7	α=0,9
1	0,48	0,38	0,50	0,62	0,74	0,61	0,48	0,34	0,25
2	0,84	0,72	0,80	0,88	0,96	0,92	0,84	0,76	0,65
3	0,48	0,46	0,50	0,54	0,58	0,52	0,48	0,44	0,41
4	0,56	0,52	0,60	0,68	0,76	0,65	0,56	0,47	0,42
5	0,56	0,52	0,60	0,68	0,76	0,64	0,56	0,48	0,43

Fig. 2 Indexes based on arithmetical mean, MOOWA and MEOWA

$y \in H_j$, with $i < j$, then element y has more importance in the aggregation process than x.

- Elements belonging to the same class should be aggregated without giving more weight to one variable than other.

In [1], a prioritized index was built based on the work [10] consisting of two steps. In a first step each a sub-index or factor was built with the information of each level. After that each level information was aggregated in an hierarchical way.

In this work, we are going to focus on the first step, in which it is necessary to build social sub-indices or factors aggregating different variables or criteria not ordered and with the same importance. Let's look at this in an example:

The intention is to make an index that measures the level of citizens confidence in politics. Obviously, an index of these characteristics will have different factors with a certain structure, usually hierarchical. Nevertheless, and independently of its structure, it is necessary to group the input information without a determined order for the different clusters that compose the index. Suppose that you have a set of citizens who have responded their degree of agreement with some sentences (1 : *strongly disagree*, ...5 : *strongly agree*), obtaining the following data (Fig. 1):

Suppose we are interested in ranking these five citizens on their level of confidence in politics. Different aggregation operators are used to comparatively analyze the resulting ranking. The standardized and aggregated information is presented in the following table (Fig. 2):

In the first place it is observed that the subject with the greatest credibility in the policy is placed in first place independently of the aggregating operator. The difference between the different operators occurs when the degrees of agreement of the same subjects are more dispersed. For example, subject number 1, in spite of neither feeling represented nor trusting politicians, considers that his vote can produce changes.In this case, the information aggregated through MOOWA places it in last place with a low ORNESS, i.e. being stricter. But as the criterion is relaxed, the value of the index increases, ascending its position. While observing the MEOWA's results, it can be seen that its values do not follow a clear trend when the ORNESS varies, alternating its position between the last and penultimate position on the level of confidence with politics. When we analyze what happens with the arithmetic mean, it can be seen that the first and third citizens of the table rank in the same position, while the sum of their valuations is the same. However, clearly the third citizen has a higher level of credibility than the first one, as he gives a greater value to *"I feel represented by some political party"*. The same applies to citizens 4 and 5.

5 Final Remarks

In this article we address a concrete problem of construction of social indexes in which the variables that are involved have the same importance. This problem arises naturally in complex hierarchical indexes where there is a need to add variables amongst which it is not possible to establish a ranking of importance in the face of aggregation. For this problem the usage of OWA aggregation operators has been proposed and within them the robustness of the MOOWA class has been observed in comparison with other techniques of weight calculation.

References

1. Rojas, K., Gómez, D., Rodríguez, J.T., Montero, J., Valdivia, A., Paiva, F.: Development of child's home environment indexes based on consistent families of aggregation operators with prioritized hierarchical information. Fuzzy Sets Syst. **241**, 41–60 (2013)
2. Yager, R.R.: On ordered weighted averaging aggregation operators in multicriteria decision-making. IEEE Trans. Syst., Man Cybern. **18**(1), 183–190 (1988)
3. Calvo, T., Mayor, G., Torrens, J., Suner, J., Mas, M., Carbonell, M.: Generation of weighting triangles associated with aggregation functions. Int. J. Uncertain. Fuzziness Knowl.-Based Syst. **8**(4), 417–451 (2000)
4. Gómez, D., Montero, J.: A discussion of aggregation functions. Kybernetika **40**, 107–120 (2004)
5. Montero, J., González-del-Campo, R., Garmendia, L., Gómez, D., Rodríguez, J.T.: Computable aggregations. Inf. Sci. **460**, 439–449 (2018)
6. Olaso, P., Rojas, K., Gómez, D., Montero, J.: A generalization of stability for families of aggregation operators. Fuzzy Sets Syst. (2019). https://doi.org/10.1016/j.fss.2019.01.004

7. O'Hagan, M.: Aggregating template or rule antecedents in real-time expert systems with fuzzy set logic. In: Proceedings of the 22nd Annual IEEE Asilomar Conference on Signals, Systems, Computers, Pacilc Grove, CA, pp. 81–89 (1988)
8. Martinez, N., Gómez, D., Olaso, P., Montero, J., Rojas, K.: A novel ordered weighted averaging weight determination based on ordinal dispersion. Int. J. Intell. Syst. **34**, 2291–2315 (2019)
9. Berry, K.J., Mielke, P.W., Jr.: Assessment of variation in ordinal data. Percept. Mot. Ski. **74**, 63–66 (1992)
10. Yager, R.R., Rybalov, A.: Noncommutative self-identity aggregation. Fuzzy Sets Syst. **85**, 73–82 (1997)

Combining Conceptual Graphs and Sentiment Analysis for Fake News Detection

Walter Cuenca, César González-Fernández, Alberto Fernández-Isabel, Isaac Martín de Diego, and Alejandro G. Martín

Abstract Misinformation has always existed in our society. In these days, the emergence of technological development and the appearance of social networks, pseudo-newspapers and blogs, has aggravated this problem by facilitating the rapid spread of fake news. This fact eases the use of misinformation as a vector of attack on huge communities, which may even compromises the security of a country. This leads to the development of systems that detect the appearance of this type of news and mitigate their influence. This article presents a first prototype of a knowledge-based system for the detection of fake news using reliable information sources. This framework makes use of text mining and sentiment analysis techniques to build conceptual graphs that represent the extracted knowledge. It can be estimated if a text provides misinformation combining similarity matrices gathered from them. Preliminary experiments confirm the potential of the proposal.

Keywords Fake news · Combination of information · Text mining · Sentiment analysis · Graph-based systems

---

W. Cuenca · C. González-Fernández · A. Fernández-Isabel (✉) · I. Martín de Diego · A. G. Martín
Universidad Rey Juan Carlos, C/Tulipán s/n, 28933 Móstoles, Madrid, Spain
e-mail: alberto.fernandez.isabel@urjc.es

W. Cuenca
e-mail: walter.cuenca@urjc.es

C. González-Fernández
e-mail: cesar.gonzalezf@urjc.es

I. Martín de Diego
e-mail: isaac.martin@urjc.es

A. G. Martín
e-mail: alejandro.garciam@urjc.es

M. E. Cornejo et al. (eds.), *Computational Intelligence and Mathematics for Tackling Complex Problems 2*, Studies in Computational Intelligence 955,
https://doi.org/10.1007/978-3-030-88817-6_15

1 Introduction

Misinformation and manipulation of information sources is a problem that has always existed in our society. The main objective is to influence communities of individuals to obtain from them a reaction to attain a benefit [17].

Nowadays, the creation of documents with fake or biased content to provoke misinformation and manipulation is a large-scale phenomenon. This is mainly due to the great technological development of telecommunications and the emergence of social networks, pseudo-newspapers and personal blogs.

The detection of documents that produce misinformation in society is a very demanding task that cannot be addressed as a manual activity. It is due to this rapid spread and the large volume of information that can be generated using the Internet as a propagation medium. Therefore, this leads to using automatic techniques able to achieve this work. Machine Learning (ML) techniques [13] appear as a core solution for this issue. These techniques allow training systems capable of detecting this kind of documents by recognizing specific patterns. The proposals in this line use techniques that can be classified into methods based on: *linguistic characteristics*, *cheating models*, *clustering*, *predictive models*, *content features* and *non-textual characteristics* [14].

This approach proposes a framework based on *content features* capable of tackling the problem of misinformation for the specific case of fake news. To this end, it uses a set of well-known web information sources to extract news which are processed through text mining techniques. This process generates a knowledge base that uses conceptual graphs. These graphs are completed through sentiment analysis techniques [8]. From these graphs, similarity matrices are obtained and combined to apply machine learning models. These models estimate the reliability of a document. The system daily updates the knowledge base, including new texts and their publisher through a time line. This allows evaluating new documents (fake or not) by contrasting them with the information stored on a specific day.

Preliminary experiments have been achieved to test the results provided by the system. In a first look, these results seem to be acceptable and confirm the potential of the applied methodology.

This paper is organized as follows. Section 2 introduces some of the most representative works in the domain. Section 3 addresses the methodology of the system. Section 4 details representative experiments of the system. Finally, Sect. 5 presents the conclusions and future lines.

2 Related Work

This section presents previous works that try to solve the misinformation problem using different ML methods. As mentioned above, due to the magnitude of the problem, the development of systems with the capacity to learn the features by studying the *content of the texts* or the *social context* has become a fundamental issue in the domain [16].

The approaches based on the *content of the texts* are focused on analyzing these texts in search of relevant concepts that allow solving the problem. In [3], a system is presented to detect the concordance between the title and the body of an article. This system focuses on the detection of fake news in the case of *Clickbait*. In [18], the *LIAR* dataset is used for the detection of fake news. It consists of an hybrid neural network that uses the information and meta-data of the text. On the other hand, [1] makes the detection of false publications on social networks. Thus, the system obtains a set of texts that are stored in a database for later use. When trying to validate a new text, those similar to it previously stored are selected. From these texts, the named entities are extracted [11], together with the core topic they are dealing with, to apply sentiment analysis techniques with the obtained information.

For the approaches based on the *social context*, they study the way of information propagates. In [9], fake news detection is performed based on the propagation structure within Twitter. Tree-like structures are used since they provide valuable clues as to how an original message is transmitted and developed over time. In [7], it is proposed to address the fake news detection by applying sentiment analysis techniques. In this case, the conflicting views existing in microblogs about news tweets are tackled.

Based on the exploration achieved on the domain, it can be indicated that the proposal of this paper is placed in the analysis of the *content of the texts*. Unlike existing approaches, this is characterized by combining different types of information and using a dynamic knowledge base against the use of static data. For the analysis of the text, the detection of relevant terms is used. It addresses the issue of ambiguity, allowing the development of a more robust system.

3 Framework Architecture

The framework performs its tasks according to two main processes: obtain information from trusted web information sources and process them, and analyze news texts to assess their misinformation level. The system has one main module that achieves both tasks (see Fig. 1). This module is completed with a knowledge base and a graphical interface.

Regarding the knowledge base structure (see Fig. 2), it uses knowledge graphs to relate the information. Thus, it defines the node *ROOT* (green node) that serves to unify the information and four other main nodes: *SOURCE* (yellow node), *DATE* (red node), *EN* (purple node) and *WORD* (blue node). The first stores the name of the web information source from which the knowledge is extracted. The second indicates the date of publication, which is useful to organize current issues according to the corresponding day. The third is responsible for storing the named entities, and finally the fourth considers the relevant words detected. The relationships between the nodes are unidirectional. Thus, the following are generated: *PUBLISHED* (between the

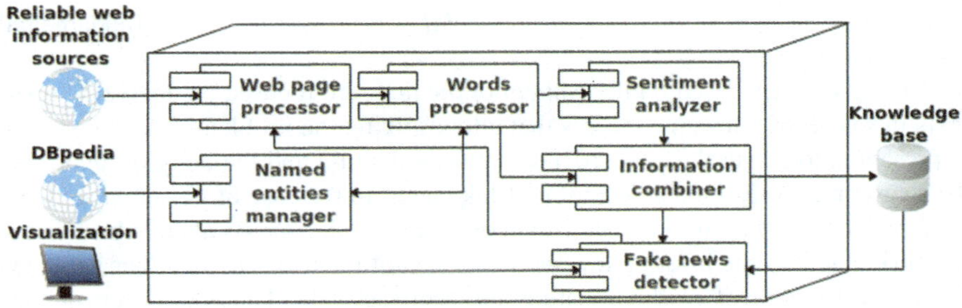

Fig. 1 Overview of the framework with the main module and components

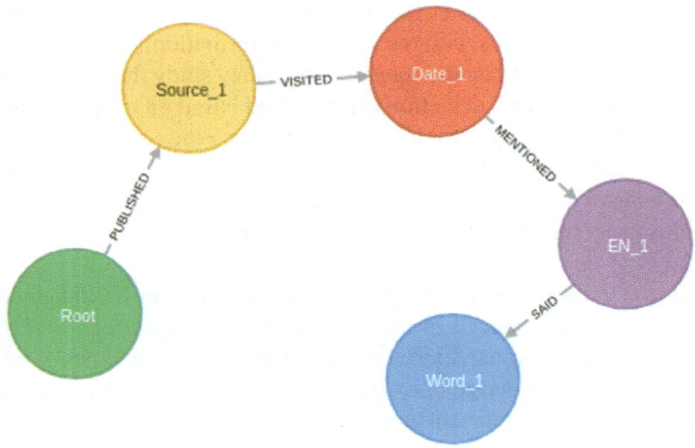

Fig. 2 Proposed knowledge base structure

nodes *ROOT* and *SOURCE* relating the name of the analyzed information source), *VISITED* (between the nodes *SOURCE* and *DATE* relating the date of publication of the text), *MENTIONED* (between the nodes *DATE* and *EN* related to the mentions detected on the date of publication), and *SAID* (between the nodes *EN* and *WORD* related to a named entity with its relevant words).

In the case of the main module, it presents six components: *Web page processor*, *Words processor*, *Named entities manager*, *Sentiment analyzer*, *Information combiner* and *Fake news detector*. All of them perform a sequential process. Note that *Fake news detector* is only used when a text is going to be evaluated and compared with the information stored in the knowledge base.

The *Web page processor* component obtains textual information through web scraping techniques [12]. This information is organized into paragraphs.

The *Words processor* component analyzes the text to detect named entities and their associated words. Thus, each paragraph is divided into words, then it applies a set of filters that obtain the grammatical category of the terms, selecting only the verbs, adjectives and nouns.

The *Named entities manager* component addresses possible problems related to the ambiguity of the language [15] (when a named entity takes different forms, e.g. *Michael Jordan* could be the same named entity as *M. Jordan* or *Jordan*). The system uses the DBpedia API REST [2] to solve this problem.

The *Sentiment analyzer* component is in charge of obtaining the polarity of the relevant words related to the named entity to obtain the sentimental value of the portions of the text (paragraphs or sentences) where it appears. The SenticNet dictionary [4] has been selected to achieve it. The sentiment value is calculated for each one of the named entities as the average sentiment value of its associated words. Notice that for more than one occurrence of the same named entity in the same text, the resulting sentiment value is the average of the associated words related to the occurrences.

The *Information combiner* component generates a matrix (M^A). The value of each one of the elements of the matrix ($M^A_{i,j}$), is the number of occurrences of the relevant word j associated to the named entity i. For each named entity, the pairs *(W,#)* are constructed, where W is the relevant word and # the number of occurrences of the named entity. Finally, a triplet *(W,#,sent)* is generated where *sent* is the associated sentiment value previously calculated. These triplets are used to produce sub-graphs which are stored in the knowledge base.

The *Fake news detector* component calculates the veracity of a text. Its validation is achieved by consulting the information available in the knowledge base. To address it, sub-graphs formed by named entities and their words for each of the sources are obtained. Two matrices of dimension SxS are generated from these sub-graphs, where S indicates the number of sources related to the text to evaluate. The first matrix (M^E) contains the similarity based on the average sentimental value between the named entities that have in common i and j sources. The second matrix (M^G) contains the similarity based on the semantic value of the words associated to the set of named entities [19]. The value of each element ($M^G_{i,j}$) is calculated as the average value of the matrix of similarity between entities according to the semantics of their words for each of the sources. Both matrices are combined through a weighted sum. A parameter called α controls the relative importance of each matrix. The resulting matrix is used to feed a veracity detection system. This system uses a *One-Class SVM* model [10] which is able to detect the anomalous texts.

4 Experiments

This section illustrates the preliminary results for testing the proposed system. Thus, two experiments are defined: *System initialization* (see Sect. 4.1) and *Evaluation of new content* (see Sect. 4.2). The first aims to evaluate how the system works when carrying out the tasks: extraction, processing and storage of the information provided by the reliable web sources. In the second, the effectiveness of the system to detect news that present possible misinformation in their content is considered.

Fig. 3 Excerpt of the sentiment result from the *BBC News* text

```
('enrich', 1.0, 0.188),
('significant', 1.0, 0.125),
('believe', 2.0, 0.1),
('complex', 2.0, -0.025),
('talk', 2.0, -0.023),
('ask', 1.0, 0.018),
('agree', 1.0, 0.232),
('return', 1.0, 0.013),
('side', 1.0, -0.009),
('relief', 1.0, 0.102),
('dependent', 1.0, -0.161),
('official', 2.0, 0.036),
('sanction', 1.0, 0.071),
('issue', 1.0, 0.008),
('exist', 1.0, 0.062),
('unspecified', 2.0, -0.625),
('contentious', 1.0, 0.375)
```

4.1 System Initialization

For this first experiment, eight web information sources have been selected labeling them as reliable: *Angeles Times*, *BBC News*, *Chicago Sun-Times*, *The Guardian*, *Economic Times*, *Independent*, *NN Politics* and *NY Post*.

Once these sources are selected, the system processes each one of them to obtain the texts of interest through the *Web page processor*. From texts, the associated named entities are gathered using the *Words processor* and the *Named entities manager* components. Next, the sentiment value of the relevant words associated with the detected named entities is calculated by the *Sentiment analyzer* component, and the corresponding triplets are produced. Finally, the information obtained is stored in the knowledge base.

For instance, let consider the news *«Trump-Kim summit: North Korea says country seeks partial relief»* from the newspaper *BBC News*. It produces a set of triplets (see Fig. 3) and a knowledge sub-graph (see Fig. 4). Notice that the green node is the central node of the graph, while the yellow node indicates the source of information processed. For the violet and blue ones, they represent the named entities and the detected relevant words respectively).

The result of the experiment shows that the system satisfactorily works. It has been able to gather and process the information from multiple web sources, generating the sub-graphs and storing them into the knowledge base.

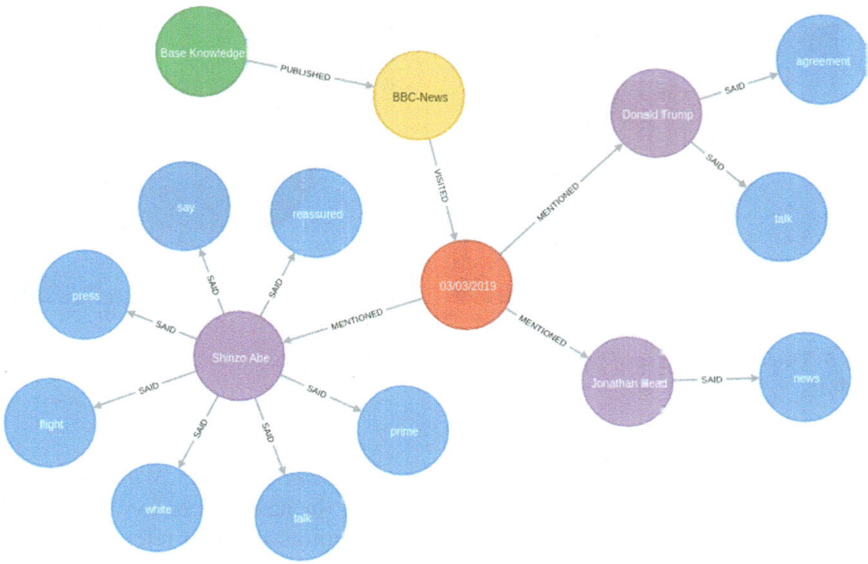

Fig. 4 Knowledge graph representation obtained from the *BBC News* text

4.2 Evaluation of New Content

In this second experiment, news with possible fake content or with misinformation are evaluated. A topic that is very relevant at the international level has been selected to tackle this issue. This selection ensures that the system has previous gathered information from the reliable web sources. The topic has been the *Hanoi summit* between the presidents of the United States and North Korea, *Donald Trump* and *Kim Jong-un* respectively.

The experiment considers two cases. In the first one, the system evaluates a real text from a trusted source previously unconsidered (i.e., the text has not been processed by the system yet). The second one consists of evaluating a synthetic text with erroneous and fake content: *«The President of the United States, Donald Trump, declared that the meeting with Kim Jon-un was very productive. Donald Trump was very happy and had the honor to indicate that the North Korean leader accepted all the proposed conditions. Kim Jong-un said there was no insecurity in the pact, since nuclear weapons will be developed according to the new agreement and relations between the two countries will improve.»*. In this way, the expected result of the experiment should be only a fake news detection in the second case, understanding the first as truthful information.

Before starting, it is necessary to define the system configuration parameters. The parameter α is set to 0.5. This means that the importance of the semantic matrix with respect to the sentiment matrix when they are combined is the same (see Sect. 3). The resulting matrix is used to train a *One-Class SVM* model.

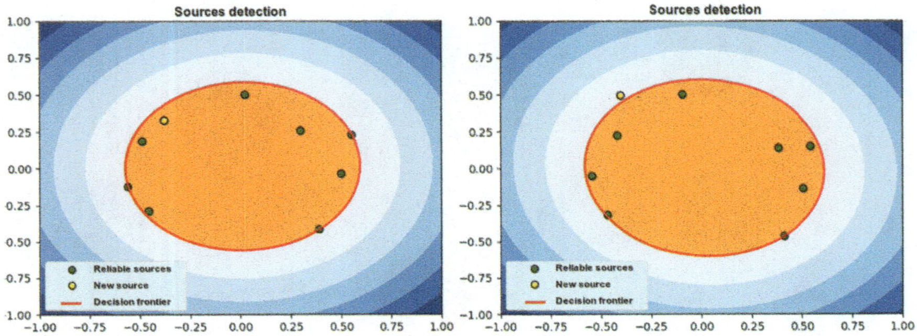

Fig. 5 Detection results for *The Washington Post* (left) and for the synthetic text with fake content (right)

In the first case, the news with title *«Trump on Kim Jong-un: 'Why should not I like him?'»* of newspaper *The Washington Post* in its web version is selected. When the content is processed, the set of named entities and their relevant words is obtained. For instance, the entity of type person *Donald Trump* has the words *news* and *conference* associated. In this case, the set of named entities are consulted in the knowledge base, obtaining that in eight of the existing sources there is at least one of the detected named entities. For example, for *The Guardian*, the named entity *Donald Trump* has the words *good* and *stuck* among others. With the information obtained from the analysis and from the knowledge base the similarity matrices are generated based on the sentiment and the semantic value of the words. In turn, these are combined by applying the value of the α parameter mentioned above.

For the second case, the text is processed in the same way as the previous one. Thus, another two similarity matrices are obtained (i.e., both the sentiment and the semantic ones).

Finally, the resulting matrix for each one of the cases are selected as an input of the *One-Class SVM* model. A Multi-Dimensional Scaling (MDS) technique [5] is applied to the resulting matrix to visualize the results. The result of this experiment shows a basic capacity of detection by the system (see Fig. 5). Thus, when a text contains similar information both at the sentiment and semantic level, the system is able to indicate the information is reliable. In contrast, if the information expresses sentiments or uses a very different content, the system is also able to detect it.

5 Conclusions

This article presents a framework to detect fake news using document content. It gathers knowledge from several web information sources through text mining and sentiment analysis techniques to build a knowledge base. This information can be combined to produce a matrix that feeds a ML model able to detect suspicious texts that could present misinformation.

Basic experiments has been carried out to show the potential of the proposed system. They have confirmed that the system is able to store properly the information. It is also able to detect fake news in a controlled environment (using synthetic texts), and detect trusted news in a real environment (using a previously unconsidered text). Though, it can be considered as a prototype tool with a complete functionality, more experiments with real data have to be accomplished. This will allow detecting possible weaknesses of the proposal and providing guidelines for the future system development.

Finally, it is desired to improve the treatment of information by building a more complete knowledge base. Possible improvements in the text processing tasks, such as the detection of co-references could substantially improve the information obtained. The use of synonyms and hypernonyms provided by tools such as WordNet [6] could also helpful.

Acknowledgements Research supported by grants from the Spanish Ministry of Economy and Competitiveness, under the Retos-Colaboración program: SABERMED (Ref: RTC-2017-6253-1); and the Council of Education, Youth and Sports of the Community of Madrid and the European Social Fund of the European Union (Ref: PEJD-2017-PRE/TIC-3901, PEJ-2017-AI/TIC-6403).

References

1. Atodiresei, C.S., Tănăselea, A., Iftene, A.: Identifying fake news and fake users on twitter. Procedia Comput. Sci. **126**, 451–461 (2018)
2. Auer, S., Bizer, C., Kobilarov, G., Lehmann, J., Cyganiak, R., Ives, Z.: Dbpedia: a nucleus for a web of open data. In: The Semantic Web, pp. 722–735. Springer (2007)
3. Bourgonje, P., Schneider, J.M., Rehm, G.: From clickbait to fake news detection: an approach based on detecting the stance of headlines to articles. In: Proceedings of the 2017 EMNLP Workshop: Natural Language Processing Meets Journalism, pp. 84–89 (2017)
4. Cambria, E., Poria, S., Bajpai, R., Schuller, B.: Senticnet 4: a semantic resource for sentiment analysis based on conceptual primitives. In: Proceedings of COLING 2016, the 26th International Conference on Computational Linguistics: Technical Papers, pp. 2666–2677 (2016)
5. Cox, T.F., Cox, M.A.: Multidimensional Scaling. Chapman and Hall/CRC (2000)
6. Fellbaum, C.: Wordnet. The Encyclopedia of Applied Linguistics (2012)
7. Jin, Z., Cao, J., Zhang, Y., Luo, J.: News verification by exploiting conflicting social viewpoints in microblogs. In: Thirtieth AAAI Conference on Artificial Intelligence (2016)
8. Liu, B.: Sentiment Analysis: Mining Opinions, Sentiments, and Emotions. Cambridge University Press (2015)
9. Ma, J., Gao, W., Wong, K.F.: Detect rumors in microblog posts using propagation structure via kernel learning. In: Proceedings of the 55th Annual Meeting of the Association for Computational Linguistics (Volume 1: Long Papers), vol. 1, pp. 708–717 (2017)
10. Manevitz, L.M., Yousef, M.: One-class SVMs for document classification. J. Mach. Learn. Res. **2**(Dec), 139–154 (2001)
11. Mansouri, A., Affendey, L.S., Mamat, A.: Named entity recognition approaches. Int. J. Comput. Sci. Netw. Secur. **8**(2), 339–344 (2008)
12. Mitchell, R.: Web Scraping with Python: Collecting More Data from the Modern Web. O'Reilly Media, Inc. (2018)

13. Mohri, M., Rostamizadeh, A., Talwalkar, A.: Foundations of Machine Learning. MIT Press (2018)
14. Parikh, S.B., Atrey, P.K.: Media-rich fake news detection: a survey. In: 2018 IEEE Conference on Multimedia Information Processing and Retrieval (MIPR), pp. 436–441. IEEE (2018)
15. Shen, W., Wang, J., Han, J.: Entity linking with a knowledge base: issues, techniques, and solutions. IEEE Trans. Knowl. Data Eng. **27**(2), 443–460 (2015)
16. Shu, K., Sliva, A., Wang, S., Tang, J., Liu, H.: Fake news detection on social media: a data mining perspective. ACM SIGKDD Explor. Newsl. **19**(1), 22–36 (2017)
17. Tudjman, M., Mikelic, N.: Information science: science about information, misinformation and disinformation. In: Proceedings of Informing Science+ Information Technology Education, pp. 1513–1527 (2003)
18. Wang, W.Y.: "liar, liar pants on fire": A new benchmark dataset for fake news detection. arXiv preprint arXiv:1705.00648 (2017)
19. Zhu, G., Iglesias, C.A.: Computing semantic similarity of concepts in knowledge graphs. IEEE Trans. Knowl. Data Eng. **29**(1), 72–85 (2017)

A Weakened Notion of Congruence to Reduce Concept Lattices

Roberto G. Aragón, Jesús Medina, and Eloísa Ramírez-Poussa

Abstract This paper addresses the problem of attribute and size reduction of concept lattices in formal concept analysis. The reduction of the number of attributes in a formal context produces a partition on the set of concepts of the concept lattice. In this work, we introduce a weaker notion of congruence relation, called local congruence. This less restrictive kind of congruence guarantees that each subset of the partition forms a closed algebraic substructure, aggregating as few concepts as possible and preserving the main information.

Keywords Formal concept analysis · Size reduction · Attribute reduction · Congruence

1 Introduction

The large amount of relevant information collected in databases have aroused the interest in developing mathematical tools to manage, obtain and treat knowledge systems. One of these tools is the mathematical theory of Formal Concept Analysis (FCA) [9]. Since real databases usually contain unnecessary information, one of the most appealing topics within FCA is related to reduction mechanisms. We can find many works which analyze methods with the goal of reducing the number of attributes in contexts [1, 2, 4–7, 10–15]. Recently, [3] has presented a mechanism

Partially supported by the State Research Agency (AEI) and the European Regional Development Fund (FEDER) project TIN2016-76653-P.

R. G. Aragón (✉) · J. Medina · E. Ramírez-Poussa
Department of Mathematics, University of Cádiz, Cádiz, Spain
e-mail: roberto.aragon@uca.es

J. Medina
e-mail: jesus.medina@uca.es

E. Ramírez-Poussa
e-mail: eloisa.ramirez@uca.es

M. E. Cornejo et al. (eds.), *Computational Intelligence and Mathematics for Tackling Complex Problems 2*, Studies in Computational Intelligence 955,
https://doi.org/10.1007/978-3-030-88817-6_16

to reduce formal contexts based on the philosophy of reduction considered in Rough Set Theory, which is a theory closely related to FCA. In this previously mentioned study, the authors showed that when we reduce the number of attributes of a context, we are inducing an equivalence relation on the set of concepts of the original concept lattice. In addition, they also proved that the structure of the resulting equivalence classes are join-semilattices. This is the main result that has motivated the study presented in this work.

In this paper, we are interested in complementing the research given in [3]. Specifically, we are interested in the study of mechanisms to reduce concept lattices by means of equivalence classes in the concepts, satisfying that these classes are closed algebraic substructures, specifically, convex sublattices of the original concept lattice, that is, subsets of concepts closed by the supremum and infimum and that any element between two concepts of a class must also be in the class. This target is achieved by the notion of congruence relation on lattices [8]. However, the reductions based on congruence relations lead to a significant loss of information, as we will show in this work. Therefore, in order to overtake this drawback, we will introduce a new and weaker notion of congruence, which we have called local congruence. We see that the obtained results from the consideration of local congruence relations improve the ones provided by congruence relations and achieve the main goal of the paper.

2 Preliminaries

First of all, we recall the basic definitions in FCA [9] needed to understand the results which have motivated the presented study. In FCA, a *context* is a triple (A, B, R) with a set of attributes A, a set of objects B and a crisp relationship $R\colon A \times B \to \{0, 1\}$, such that $R(a, x) = aRx = 1$, if $a \in A$ and $x \in B$ are related, and $R(a, x) = 0$, otherwise. In addition, considering a context, two mappings, ${}^{\uparrow}\colon 2^B \to 2^A$ and ${}^{\downarrow}\colon 2^A \to 2^B$, can be defined for each $X \subseteq B$ and $Y \subseteq A$ as:

$$X^{\uparrow} = \{a \in A \mid \text{for all } x \in X, aRx\} \tag{1}$$

$$Y^{\downarrow} = \{x \in B \mid \text{for all } a \in Y, aRx\} \tag{2}$$

These operators form a Galois connection [8]. Therefore, if the equalities $X^{\uparrow} = Y$ and $Y^{\downarrow} = X$ hold, for a pair (X, Y), with $X \subseteq B$ and $Y \subseteq A$, this pair is called *concept*. For each pair of concepts $(X_1, Y_1), (X_2, Y_2) \in \mathcal{C}(A, B, R)$, if $X_1 \subseteq X_2$, then $(X_1, Y_1) \leq (X_2, Y_2)$.

The set of all the concepts with the previously defined ordering relation has the structure of a complete lattice, is called *formal concept lattice* and it is denoted as $\mathcal{C}(A, B, R)$ [8, 9]. The next result shows that when we reduce the set of attribute, an equivalence relation on the set of concepts of the original concept lattice is induced.

Proposition 1 ([3]) *Given a context* (A, B, R) *and a subset* $D \subseteq A$. *The set* $R_E = \{((X_1, Y_1), (X_2, Y_2)) \mid (X_1, Y_1), (X_2, Y_2) \in \mathcal{C}(A, B, R), X_1^{\uparrow_D\downarrow} = X_2^{\uparrow_D\downarrow}\}$ *is an* *equivalence relation. Where* $\uparrow_D$ *denotes the concept-forming operator, given in Expression* (1), *restricted to the subset of attributes* $D \subseteq A$.

Every class of the previously defined equivalence relation is a join semilattice with maximum element, as the following result shows.

Proposition 2 ([3]) *Given a context* (A, B, R), *a subset* $D \subseteq A$ *and a class* $[(X, Y)]_D$ *of the quotient set* $\mathcal{C}(A, B, R)/R_E$. *The class* $[(X, Y)]_D$ *is a join semilattice with maximum element* $(X^{\uparrow_D\downarrow}, X^{\uparrow_D\downarrow\uparrow})$.

Now, we will introduce the notion of congruence on a lattice and some properties which are essential to develop our work. First of all, before introducing the notion of congruence we need to recall the following definition.

Definition 1 We say that an equivalence relation θ on a given lattice $(L, \preceq)$ is *compatible* with join and meet if, for all $a, b, c, d \in L$,

$$a \equiv b \ (\text{mod}\ \theta) \ \text{ and } \ c \equiv d \ (\text{mod}\ \theta)$$

imply $a \vee c \equiv b \vee d \ (\text{mod}\ \theta)$ and $a \wedge c \equiv b \wedge d \ (\text{mod}\ \theta)$.

Below we recall the definition of congruence on a lattice.

Definition 2 Given a lattice $(L, \preceq)$, we say that an equivalence relation on L, which is compatible with both join and meet is a *congruence* on L.

We will write $a \equiv b \ (\text{mod}\ \theta)$ or $(a, b) \in \theta$ to indicate that a and b are related under the congruence θ. Given a lattice L and a subset of L composed of four elements $\{a, b, c, d\}$, if $a < b$, $c < d$ and either ($a \vee d = b$ and $a \wedge d = c$) or ($b \vee c = d$ and $b \wedge c = a$), then a, b and c, d are said to be *opposite sides of the quadrilateral*. In addition, we say that the equivalence classes provided by a congruence are *quadrilateral-closed* if whenever given two opposite sides of a quadrilateral a, b and c, d, satisfying that a, b belong to one equivalence class X, then $c, d \in Y$ for another equivalence class Y.

Theorem 1 ([8]) *Let* $(L, \preceq)$ *be a lattice and let* θ *be an equivalence relation on* L. *Then* θ *is a congruence if and only if:*

(i) each equivalence class of θ *is a sublattice of* L,
(ii) each equivalence class of θ *is convex,*
(iii) the equivalence classes of θ *are quadrilateral-closed.*

The set of congruences on a lattice L, denoted as $\text{Con}\, L$, is a topped $\cap$-structure on $L \times L$. Hence $\text{Con}\, L$, together with the inclusion ordering, is a complete lattice. The bottom element and the top element are given by $\{(a, a) \mid a \in L\}$ and $\{(a, b) \mid a, b \in L\}$, respectively.

Table 1 Relation of Example 1

R	M	V	E	Ma	J	S	U	N	P
small size	1	1	1	1	0	0	0	0	1
medium size	0	0	0	0	0	0	1	1	0
large size	0	0	0	0	1	1	0	0	0
near sun	1	1	1	1	0	0	0	0	0
far sun	0	0	0	0	1	1	1	1	1
moon yes	0	0	1	1	1	1	1	1	1
moon no	1	1	0	0	0	0	0	0	0

3 Local congruences

As we mentioned previously, reductions of attributes in FCA generate a partition in the set of concepts of the original context whose equivalence classes may not necessarily form sublattices. At certain times, it could be interesting that the reduction generates groups of concepts with the structure of sublattices in order to have more homogeneous classes with respect to the ordering among the concepts and so, on a complete algebraic structure. In this section, we will see how we can complete the reduction given in FCA to satisfy the mentioned structural property.

First of all, we show the result of applying congruence relations on the concept lattice, in a practical example. In particular, we want to find the smallest congruence that contains the equivalence classes of the reduction obtained from FCA. We illustrate this idea by means of the following example, which was considered in [3].

Example 1 We consider the formal context (A, B, R) displayed in Table 1, where the set of objects in B are the planets of the Solar System together with the dwarf planet Pluto, that is $B = \{$Mercury (M), Venus (V), Earth (E), Mars (Ma), Jupiter (J), Saturn (S), Uranus (U), Neptune (N), Pluto (P)$\}$ and the set of attributes $A = \{$small size (ss), medium size (ms), large size (ls), near sun (ns), far sun, (fs), moon yes (my), moon no(mn)$\}$.

The concept lattice obtained from this context is given in the left side of Fig. 1. In [3], following the steps of reduction mechanism, it was carried out a reduction from the subset of attributes $D_1 = \{$small size, medium size, near sun, moon yes$\}$. Using such a reduction, the concepts of the original concept lattice are grouped in equivalence classes which are represented in the middle of Fig. 1. As it was mentioned in Proposition 2, each equivalence class has the structure of a join semilattice with maximum element.

We will show how we can bring together the reduction given in FCA and congruences. We find the smallest congruence such that each equivalence class induced by the reduction of the context, is included in one equivalence class provided by the congruence relation. This congruence is shown in the right side of Fig. 1. As we can

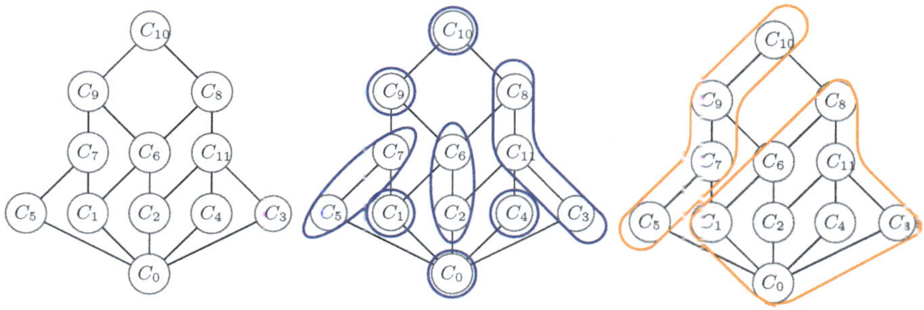

Fig. 1 The original concept lattice (left), the obtained reduction in [3] (middle) and the smallest congruence containing the previously reduction (right)

see in Fig. 1, this congruence relation has grouped many concepts in the same class and, consequently a lot of information is lost.

The result obtained in the previous example highlights the necessity of a weaker notion of congruence. This novel notion is presented in the following definition.

Definition 3 Given a lattice $(L, \preceq)$, we say that an equivalence relation δ on L is a *local congruence* if

(i) each equivalence class of δ is a sublattice of L,
(ii) each equivalence class of δ is convex.

Note that Definition 3 is similar to Theorem 1 by removing its third condition. Since δ is an equivalence relation, this definition provides a partition on L.

Definition 4 Let $(L, \preceq)$ be a lattice and δ a local congruence, the quotient set L/δ provides a partition of L, which is called *local congruence partition* (or *lc-partition* in short) of L and it is denoted as π_δ. The elements in the lc-partition π_δ are convex sublattices of L.

The definition of local congruence makes that the equivalence classes of the quotient set L/δ are closed algebraic structures. A direct characterization of the notion of local congruence in terms of the equivalence relation δ, is shown in the following result.

Proposition 3 *Given a lattice $(L, \preceq)$ and an equivalence relation δ on L, the relation δ is a local congruence on L if and only if, for each $a, b, c \in L$, the following properties hold:*

(i) If $(a, b) \in \delta$ and $a \preceq c \preceq b$, then $(a, c) \in \delta$.
(ii) $(a, b) \in \delta$ if and only if $(a \wedge b, a \vee b) \in \delta$.

Next, we define the inclusion of two local congruences taking into account the associated quotient sets.

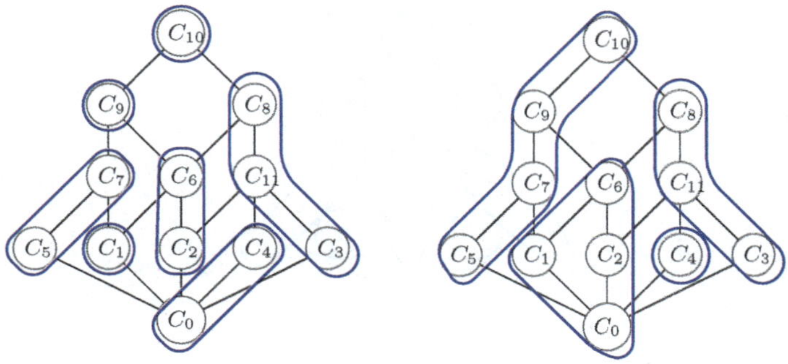

Fig. 2 Two local congruences containing the reduction from Fig. 1

Definition 5 Let $(L, \preceq)$ be a lattice and δ_1, δ_2 two local congruences on L. We say that the local congruence δ_1 is less than δ_2, denoted as $\delta_1 \sqsubseteq \delta_2$, if for every equivalence class $X \in L/\delta_1$ there exists an equivalence class $Y \in L/\delta_2$ such that $X \subseteq Y$.

All the previous notions and results are very important in order to obtain our main goal of this paper, that is, to find out the best reduction given by aggregating the concepts through closed sublattices of the original concept lattice, loosing as little information as possible. Hence, from the partition given by the reduction obtained from [3] in join-semilattices, we compute the infimum of the local congruences greater that it and so, we obtain the smaller partition on sublattices of L containing the given one.

Example 2 Returning to Example 1, there exist different local congruences containing the partition given by the reduction obtained from [3]. For example, Fig. 2 presents two of them which are incomparable and provide more granular partitions (are lesser than) than the one given by the usual definition of congruence (right diagram in Fig. 1). Hence, they offer a better reduction than the one provided using congruences, aggregating as less information as possible. Since this reduction already produces sublattices, the least local congruence containing it is itself. Thus, we can assert that the use of local congruences is the best notion to capture the main goal of the paper.

4 Conclusions and Future Work

In this work, we introduce a weaker definition of congruence relation which can be applied to the problem of attribute and size reduction in FCA. Its main properties are analyzed, highlighting that the elements in the lc-partition are convex sublattices of the original concept lattice This study will be continued in the future including complementary properties.

References

1. Alcalde, C., Burusco, A.: Study of the relevance of objects and attributes of l-fuzzy contexts using overlap indexes. In: Medina, J., Ojeda-Aciego, M., Verdegay, J.L., Pelta, D.A., Cabrera, I.P., Bouchon-Meunier, B., Yager, R.R. (eds.) Information Processing and Management of Uncertainty in Knowledge-Based Systems. Theory and Foundations. pp. 537–548. Springer, Cham (2018)
2. Antoni, L., Krajči, S., Krídlo, O.: Constraint heterogeneous concept lattices and concept lattices with heterogeneous hedges. Fuzzy Sets Syst. **303**, 21–37 (2016)
3. Benítez-Caballero, M.J., Medina, J., Ramírez-Poussa, E., Ślęzak, D.: A computational procedure for variable selection preserving different initial conditions. Int. J. Comput. Math. **0**(0), 1–18 (2019)
4. Chen, J., Mi, J., Xie, B., Lin, Y.: A fast attribute reduction method for large formal decision contexts. Int. J. Approx. Reason. **106**, 1–17 (2019)
5. Cornejo, M.E., Medina, J., Ramírez-Poussa, E.: Attribute reduction in multi-adjoint concept lattices. Inf. Sci. **294**, 41–56 (2015)
6. Cornejo, M.E., Medina, J., Ramírez-Poussa, E.: Attribute and size reduction mechanisms in multi-adjoint concept lattices. J. Comput. Appl. Math. **318**, 388–402 (2017). Computational and Mathematical Methods in Science and Engineering CMMSE-2015
7. Cornejo, M.E., Medina, J., Ramírez-Poussa, E.: Characterizing reducts in multi-adjoint concept lattices. Inf. Sci. **422**, 364–376 (2018)
8. Davey, B., Priestley, H.: Introduction to Lattices and Order, 2nd edn. Cambridge University Press (2002)
9. Ganter, B., Wille, R.: Formal Concept Analysis: Mathematical Foundation. Springer (1999)
10. Konecny, J., Krajča, P.: On attribute reduction in concept lattices: the polynomial time discernibility matrix-based method becomes the CR-method. Inf. Sci. **491**, 48–62 (2019)
11. Li, J., Kumar, C.A., Mei, C., Wang, X.: Comparison of reduction in formal decision contexts. Int. J. Approx. Reason. **80**, 100–122 (2017)
12. Liñeiro-Barea, V., Medina, J., Medina-Bulo, I.: Generating fuzzy attribute rules via fuzzy formal concept analysis. In: Kóczy, L.T., Medina, J. (eds.) Interactions Between Computational Intelligence and Mathematics. Studies in Computational Intelligence, pp. 105–119. Springer, Cham (2018)
13. Medina, J.: Relating attribute reduction in formal, object-oriented and property-oriented concept lattices. Comput. Math. Appl. **64**(6), 1992–2002 (2012)
14. Ren, R., Wei, L.: The attribute reductions of three-way concept lattices. Know.-Based Syst. **99**(C), 92–102 (2016)
15. Shao, M.-W., Li, K.-W.: Attribute reduction in generalized one-sided formal contexts. Inf. Sci. **378**, 317–327 (2017)

Interactive Search by Using Minimal Generators

P. Cordero, M. Enciso, A. Mora, M. Ojeda-Aciego, and C. Rossi

Abstract If-then rules are frequently used as basic elements for knowledge representation in several areas. In Formal Concept Analysis, these rules are the so-called *implications* and can be used to find minimal generators in a symbolic way by using logic. The computation of all minimal generators is exponential. Here, we provide a novel lazy algorithm with polynomial delay in which minimal generators are used as forks in a map to guide an interactive search.

Keywords Formal concept analysis · Attribute implications · Minimal generators · Interactive search

1 Introduction

We continue our research line focused on the development of several logic-based algorithms [1, 3, 6, 7] to efficiently manage attribute implications within Formal Concept Analysis (FCA). The general aim of this paper is to introduce a logic-guided

---

Partially supported by projects TIN15-70266-C2-P-1, TIN2017-89023-P, and PGC2018-095869-B-I00 of the Science and Innovation Ministry of Spain, co-funded by the European Regional Development (ERDF).

P. Cordero · M. Enciso · A. Mora · M. Ojeda-Aciego (✉) · C. Rossi
Universidad de Málaga, Málaga, Spain
e-mail: aciego@uma.es

P. Cordero
e-mail: pcordero@uma.es

M. Enciso
e-mail: enciso@uma.es

A. Mora
e-mail: amora@uma.es

C. Rossi
e-mail: rossi@uma.es

M. E. Cornejo et al. (eds.), *Computational Intelligence and Mathematics for Tackling Complex Problems 2*, Studies in Computational Intelligence 955,
https://doi.org/10.1007/978-3-030-88817-6_17

method to efficiently search into a concept lattice with the final goal of building an interactive filtering process following the user's preferences.

This problem is not new, and some approaches to this problem can be found in the literature. For instance, Zou et al. [11] propose a personalized recommendation system based on FCA divided into two parts: the first one in which the formal context and the concept lattice are constructed from the database, and the association rules are extracted and stored; in the second part, the preferences and the rules library are used to calculate the ordered recommendation results. Senatore and Pasi [10] provide another approach, focused on collaborative filtering, in the form of a theoretical framework suitable to generate correlations among data through a lattice design; specifically, a fuzzy annotation of the lattice allows discovering similarities among items as well as users, arranged as a ranked list.

In this work, we follow a different approach, and propose a new logic-guided implication-based algorithm to infer the minimal generators of closed attribute sets. The computation of all minimal generators is a hard problem due to its exponential nature. We face this inherent difficulty by providing a lazy algorithm with polynomial delay, in which a choice of minimal generators are used as the forks in a map to guide an interactive search.

Our approach uses the background knowledge of the dataset and the user preferences, following a conversational paradigm for decision making. The conversation starts when the user makes an initial selection of desired attributes and, then, the system obtains (in polynomial time) the set of attributes (minimal generators) which cover the user's selection. In subsequent steps, the user chooses among the minimal generators until a *formal concept* which covers all the choices made during the conversation is accepted by te user.

2 Preliminaries

We recall below some basic notions in FCA; for more details see [4].

A *formal context* is a triplet $\mathbb{K} = \langle G, M, I \rangle$ where G and M are non-empty finite sets and $I \subseteq G \times M$ is a binary relation between G and M. The elements in G and M are called objects and attributes, respectively.

Each formal context $\mathbb{K}$ defines two derivation operators, namely

$$(-)' \colon 2^G \to 2^M \text{ where } A' = \{m \in M \mid (g, m) \in I \text{ for all } g \in A\}.$$
$$(-)' \colon 2^M \to 2^G \text{ where } B' = \{g \in G \mid (g, m) \in I \text{ for all } m \in B\}.$$

which form a Galois connection between $\langle 2^G, \subseteq \rangle$ and $\langle 2^M, \subseteq \rangle$. For each $A \subseteq G$, A' is the set of attributes common to the objects in A and, if B is an attribute set, B' is the set of objects which have all the attributes in B.

Due to the fact that this pair of operators is a Galois connection, the composition of them, in any direction, is a closure operator and their fixpoints lead to the notions of formal concept and *concept lattice*. The number of elements in this lattice can be exponential with respect to the context size.

The main aim of this area is to extract knowledge from the context allowing to reason about them. One of the ways to represent knowledge is by means of the concept lattice. Another equivalent alternative knowledge representation, more suitable to define reasoning methods, is given in terms of attribute implications.

Definition 1 *(Attribute implication)* Given a formal context $\mathbb{K}$, an attribute implication is an expression $A \to B$ where $A, B \subseteq M$ and we say that $A \to B$ holds in $\mathbb{K}$ whenever $A' \subseteq B'$.

The most reasonable way to handle attribute implications is in terms of logic, and we will work with the Simplification Logic $\mathbf{SL}_{FD}$, which we briefly recall below (further details and proofs can be found in [8]).

Definition 2 *(Syntax & Semantics)*

- Given a non-empty finite alphabet S, the language of $\mathbf{SL}_{FD}$ is $\mathcal{L}_S = \{A \to B \mid A, B \subseteq S\}$. Elements in S are named attributes.
- A context $\mathbb{K} = \langle G, M, I \rangle$ is an interpretation for $\mathcal{L}_S$ when $M = S$. A model for $A \to B \in \mathcal{L}_S$ is an interpretation $\mathbb{K}$ for $\mathcal{L}_S$ such that $A' \subseteq B'$. It is denoted by $\mathbb{K} \models A \to B$.

As usual, for a context $\mathbb{K}$ and an implicational system Σ, $\mathbb{K} \models \Sigma$ means that $\mathbb{K}$ is a model for every implication in Σ and $\Sigma \models A \to B$ denotes that every model for Σ is also a model for $A \to B$. In addition, two implicational systems Σ_1 and Σ_2 are said to be equivalent, denoted by $\Sigma_1 \equiv \Sigma_2$, when the models of both implicational systems coincide (i.e. $\mathbb{K} \models \Sigma_1$ iff $\mathbb{K} \models \Sigma_2$).

We introduce the axiom system of $\mathbf{SL}_{FD}$ in which, in order to distinguish between language and metalanguage inside implications, AB means $A \cup B$ and $A \smallsetminus B$ denotes the set difference.

Definition 3 *(Axiom system)* $\mathbf{SL}_{FD}$ considers the reflexivity axiom $\vdash A \to A$; together with the following inference rules, which are named fragmentation, composition and simplification respectively.

$\{A \to BC\} \vdash A \to B$;
$\{A \to B,\ C \to D\} \vdash AC \to BD$;
If $A \subseteq C$ and $A \cap B = \varnothing$ then $\{A \to B,\ C \to D\} \vdash C \smallsetminus B \to D \smallsetminus B$.

We use the standard notation in such way that $\Sigma \vdash A \to B$ denotes that $A \to B$ can be derived or inferred from Σ by using the axiomatic system.

The above axiom system is sound and complete [8] (i.e. $\Sigma \models A \to B$ iff $\Sigma \vdash A \to B$). The main advantage of $\mathbf{SL}_{FD}$ is that its inference rules may be considered equivalence rules and they are enough to compute all the derivations. These equivalences have opened the door to the development of several automated methods to manage implications by using Simplification Logic, for instance: to compute the syntactic attribute closure [8], to reduce the specification of implications [3], to compute several kinds of basis of implications [6, 7] and, the one that has inspired this work, to compute all minimal generators and closed sets [9].

In [1] we devised a method to enumerate all minimal generators. Their inherent order structure allows for managing information similar to that provided by a concept lattice, but in a more succinct form. Concept lattices hierarchically present how objects are grouped together, characterized by a set of common attributes, corresponding with an closed set of attributes. These closed sets can also be identified in the ordered structure of minimal generators, allowing a small representation to be used as the core of our method.

3 Using Minimal Generators as Forks

The closure of a set of attributes which is strongly related with the closure in the concept lattice, by the soundness and completeness theorem, is given below.

Definition 4 Given $\Sigma \subseteq \mathcal{L}_S$ and $X \subseteq S$, the (syntactic) closure of X wrt Σ is the largest subset of S, denoted X^+_Σ, such that $\Sigma \vdash X \to X^+_\Sigma$.

Our contribution is strongly based on the Deduction Theorem ($\Sigma \vdash A \to B$ iff $\{\varnothing \to A\} \cup \Sigma \vdash \varnothing \to B$) and the following three logic equivalences.

Proposition 1 *For all $A, B, C \in M$, the following equivalences hold:*

- **Eq. I**: *If $B \subseteq A$ then $\{\emptyset \to A, B \to C\} \equiv \{\emptyset \to A \cup C\}$.*
- **Eq. II**: *If $C \subseteq A$ then $\{\emptyset \to A, B \to C\} \equiv \{\emptyset \to A\}$.*
- **Eq. III**: *Otherwise $\{\emptyset \to A, B \to C\} \equiv \{\emptyset \to A, B \smallsetminus A \to C \smallsetminus A\}$.*

An algorithm for computing A^+_Σ which, in addition, provides a reduced set of implications that can be seen as the knowledge in the system beyond this closure was given in [8]. Later, in [2], we presented how $\mathbf{SL}_{FD}$ can be used as a basis to enumerate all minimal generators and to manage social network information.

As stated in the introduction, we will use minimal generators as bifurcation points that are computed on demand in polynomial time, and presented to the user to guide his/her search. Such a method uses Simplification Logic, and is implemented in terms of the function NextMinGen.

Function NextMinGen(X,Σ)

input : X, A set of attributes selected by the user;
Σ, a set of implications describing the system knowledge;
output: $Next$, a set of attribute sets which are the choices of the next step in the guided search;
Γ, a reduced set of implications;

```
repeat
    Γ := Σ; Σ := ∅
    foreach A → B ∈ Γ do
        if A ⊆ X then X := X ∪ B
        else if B ⊈ X then Σ := Σ ∪ {A ∖ X → B ∖ X}
until Γ = Σ
Next := Minimals{A ⊆ M | A → B ∈ Γ for some B ⊆ M}
return (Next, Γ)
```

The workflow of the algorithm is described in Fig. 1. It starts by inferring the implicational system Σ from the formal context; the method to infer Σ from the dataset uses the ARules package presented in [5]. Then the user selects a set of attributes representing the initial needs. Given this user selection, our algorithm adds the formula $\varnothing \to A$ to the implicational set Σ and uses the function NextMinGen to provide, in each step, several sets of attributes, each one corresponding to a successful way to reach a solution for his previous search. The user can decide to stop the algorithm when the set of objects that have been reached is manageable; i.e. when the query to the data fulfilling all the selected attributes returns a number of items suitable for his/her needs. Otherwise, the algorithm requires the user to make a new choice and goes on until a minimal generator for the whole set of attributes is reached.

4 Conclusions and Future Works

The main contribution of this work is a new algorithm to infer minimal generators of closed attribute sets. The algorithm is built on $\mathbf{SL}_{FD}$ logic, with FCA as its framework. The main novelty of the algorithm is that it follows a lazy strategy, in such a way that its application generates processes structured in two phases. The first one is a pre-processing step with exponential complexity, that is executed only once as an initial step. The second (and iterative) phase takes polynomial time, what makes it suitable for real-world applications.

This decomposition in two phases is not rare in recommender systems [11]. Collaborative recommender systems deployed in large companies use to isolate costly processes that are executed in batch-mode (at night, for example). This is the case of updating the top-n neighbours of each user (or the distance between users) after recording daily transactions in large datasets.

Our algorithm provides the theoretical basis of systems that require an interactive searching in high-dimensional datasets (i.e., datasets with a large number of attributes). In this sense, our work allows for implementing "smart filtering" processes, in such a way that it is only required that the user provides information about the desired values for certain "dominant" attributes (which determine the value of the "dominated" ones). More formally, the dominant attributes are those included in the minimal generators calculated by our algorithm.

One of the main advantages of our approach is that it allows to shorten the length of the dialogue with the user, avoiding the supply of superfluous information in the filtering process. Besides, our method is effective and efficient. The effectiveness comes from working with implications, what ensures that the items resulting from the filtering satisfies user requirements (unlike some recommendation systems that use prediction techniques). As already mentioned, our algorithm is efficient "in practice" because of the polynomial complexity of the interactive phase.

Currently, we are testing an implementation of the system. As future work, we are working in the algorithm improvement by enriching the information of the minimal generators. Our goal is to define utility functions that allows to devise a "smarter"

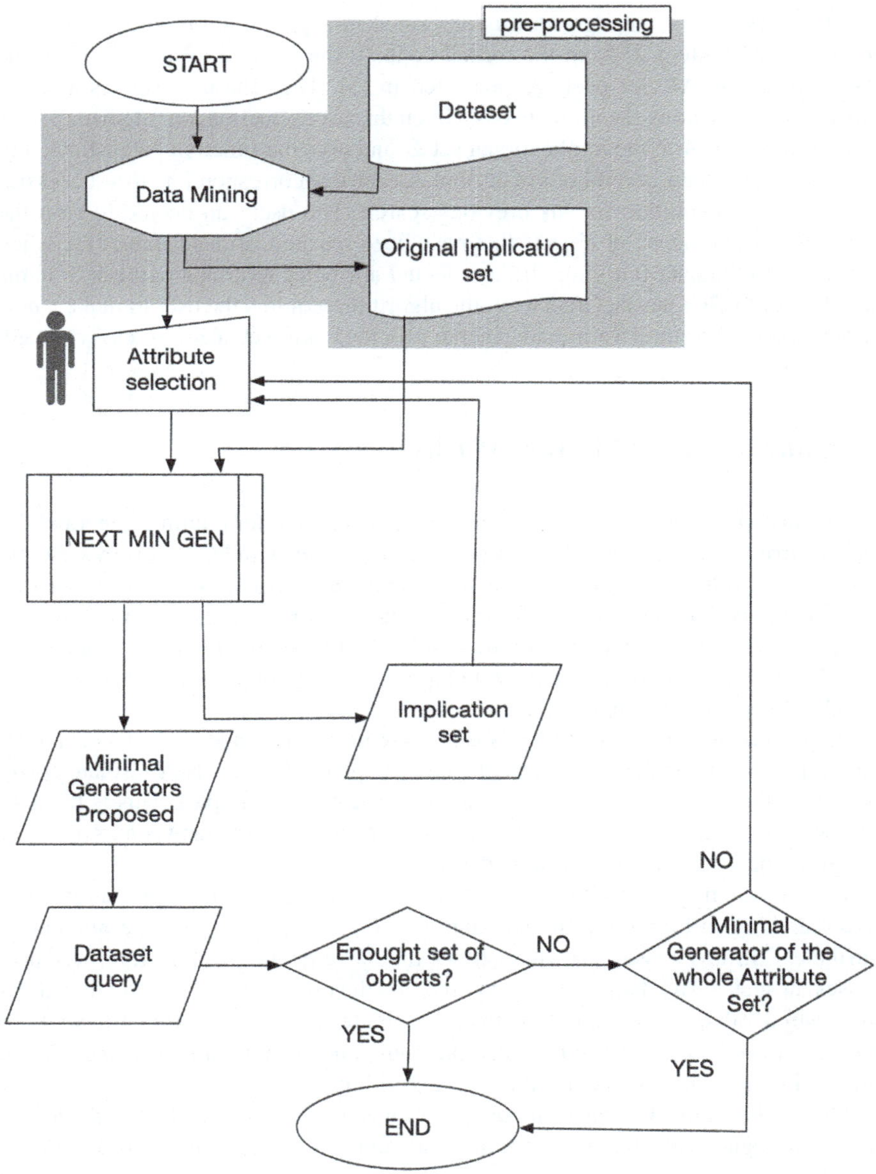

Fig. 1 Workflow of the algorithm

filtering. The idea is to discard automatically the choices that does not report an added value to the user (in the terms modeled by the corresponding utility function).

References

1. Cordero, P., Enciso, M., Mora, A., Ojeda-Aciego, M.: Computing minimal generators from implications: a logic-guided approach. In: Proceedings of 9th International Conference on Concept Lattices and Their Applications (CLA), CEUR Workshop Proceedings 972, pp. 187–198 (2012)
2. Cordero, P., Enciso, M., Mora, A., Ojeda-Aciego, M., Rossi, C.: Knowledge discovery in social networks by using a logic-based treatment of implications. Knowl. Based Syst. **87**, 16–25 (2015)
3. Cordero, P., Enciso, M., Mora, A., Rodríguez-Jiménez, J.M.: Computing non-redundant sets of functional dependencies via simplification. In: IEEE Symposium on Foundations of Computational Intelligence, FOCI, pp. 9–14. IEEE (2013)
4. Ganter, B., Wille, R.: Formal Concept Analysis: Mathematical Foundations. Springer (1999)
5. Hahsler, M., Grun, B., Hornik, K.: ARules–a computational environment for mining association rules and frequent item sets. J. Stat. Softw. **14**, 1–25 (2005)
6. Lorenzo, E.R., Adaricheva, K.V., Cordero, P., Enciso, M., Mora, A. Formation of the d-basis from implicational systems using simplification logic. Int. J. Gen. Syst. **46**(5), 547–568 (2017)
7. Lorenzo, E.R., Bertet, K., Cordero, P., Enciso, M., Mora, A.: Direct-optimal basis computation by means of the fusion of simplification rules. Discrete Appl. Math. **249**, 106–119 (2018)
8. Mora, A., Cordero, P., Enciso, M., Fortes, I., Aguilera, G.: Closure via functional dependence simplification. Int. J. Comput. Math. **89**(4), 510–526 (2012)
9. Picazo, F.B., Cordero, P., Enciso, M., Mora, A.: Minimal generators, an affordable approach by means of massive computation. J. Supercomput. **75**(3), 1350–1367 (2019)
10. Senatore, S., Pasi, G.: Lattice navigation for collaborative filtering by means of (fuzzy) formal concept analysis. In: Proceedings of the 28th Annual ACM Symposium on Applied Computing, SAC, pp. 920–926 (2013)
11. Zou, C., Zhang, D., Wan, J., Hassan, M.M., Lloret, J.: Using concept lattice for personalized recommendation system design. IEEE Syst. J. **11**(1), 305–314 (2017)

Knowledge Implications in Multi-adjoint Concept Lattices

Jesús Medina, Pablo Navareño, and Eloísa Ramírez-Poussa

Abstract In this paper, an alternative definition of implication between attributes in formal concept analysis, within the fuzzy environment of the multi-adjoint concept lattices, is presented. This novel definition establishes a relationship between crisp subsets of attributes, according to the information that these subsets provide. One of the main interests of this new definition is its potential application in the detection of unnecessary data as well as in tasks related to attribute reduction.

Keywords Knowledge implication · Formal concept analysis · Attribute reduction · Multi-adjoint concept lattice

1 Introduction

To extract information from databases is a very appealing topic in research areas as diverse as medical diagnosis, market prediction or forensic analysis. In many cases, there exist dependencies among data contained in databases. This fact may be due to the information provided by certain data can be inferred from other data. In this paper,

Partially supported by the 2014-2020 ERDF Operational Programme in collaboration with the State Research Agency (AEI) in projects TIN2016-76653-P and PID2019-108991GB-I00, and with the Department of Economy, Knowledge, Business and University of the Regional Government of Andalusia in project FEDER-UCA18-108612, and by the European Cooperation in Science & Technology (COST) Action CA17124.

J. Medina · P. Navareño · E. Ramírez-Poussa (✉)
Department of Mathematics, University of Cádiz, Cádiz, Spain
e-mail: eloisa.ramirez@uca.es

J. Medina
e-mail: jesus.medina@uca.es

P. Navareño
e-mail: pablo.navareno@uca.es

M. E. Cornejo et al. (eds.), *Computational Intelligence and Mathematics for Tackling Complex Problems 2*, Studies in Computational Intelligence 955,
https://doi.org/10.1007/978-3-030-88817-6_18

we address this specific topic considering Formal Concept Analysis (FCA) [8], one interesting mathematical tool used to extract information from relational databases.

FCA deals with databases composed of attributes and objects related between them. The information obtained from a context is represented by means of a mathematical structure, called concept lattice, which is very useful to analyze the obtained information as well as the existing dependencies among the discovered knowledge. Specifically, this work is focused on the use of this formal tool within the paradigm of the multi-adjoint concept lattice [11], due to the great potential and flexibility provided by this framework.

On the other hand, there exist different techniques devoted to obtain information contained in a context. One of the most widely studied techniques for knowledge acquisition in FCA is attribute implications, there exist papers that address this topic in the classical case [1, 13] as well as in the fuzzy one [2, 3, 9, 10, 12, 14]. In this work, we present an alternative definition of implication between attributes. This novel definition considers a different philosophy from the usual one since we do not establish a direct relation among the attributes, but between the knowledge we can obtain from these attributes. Specifically, this new implication is built from the perspective of the knowledge provided by the concept lattice, therefore, it can be understood as a knowledge implication. In this respect, a subset of attributes A_1 implies to another subset A_2 when the knowledge provided by the attributes in A_2 can be deduced from the attributes in A_1. Therefore, it can be considered as an alternative and complementary definition of implication, which can be really useful to reduce the size of the original database, facilitating the detection of unnecessary data.

2 Preliminaries

In this section we recall the preliminary needed notions to make this paper as self-contained as possible. In order to consider FCA within the multi-adjoint environment, we need to fix a multi-adjoint frame and a multi-adjoint context, we recall this notions below.

Definition 1 A *multi-adjoint frame* is a tuple $(L_1, L_2, P, \&_1, \ldots, \&_n)$ where $(L_1, \preceq_1)$ and $(L_2, \preceq_2)$ are complete lattices, $(P, \leq)$ is a poset and $(\&_i, \swarrow^i, \nwarrow_i)$ is an adjoint triple with respect to L_1, L_2, P, for all $i \in \{1, \ldots, n\}$.

More details about adjoint triples can be seen in [5, 6].

Definition 2 Let $(L_1, L_2, P, \&_1, \ldots, \&_n)$ be a multi-adjoint frame, a *context* is a tuple (A, B, R, σ) such that A and B are nonempty sets (usually interpreted as attributes and objects, respectively), R is a P-fuzzy relation $R: A \times B \to P$ and $\sigma: A \times B \to \{1, \ldots, n\}$ is a mapping which associates any element in $A \times B$ with some particular adjoint triple in the frame.

Once a multi-adjoint frame and a multi-adjoint context are fixed, the concept-forming operators $^\uparrow: L_2^B \to L_1^A$ and $^\downarrow: L_1^A \to L_2^B$ are defined as:

$$g^{\uparrow}(a) = \inf\{R(a,b) \swarrow^{a,b} g(b) \mid b \in B\} \tag{1}$$

$$f^{\downarrow}(b) = \inf\{R(a,b) \nwarrow_{a,b} f(a) \mid a \in A\} \tag{2}$$

for all $g \in L_2^B$, $f \in L_1^A$ and $a \in A$, $b \in B$, where L_2^B and L_1^A denote the set of mappings $g\colon B \to L_2$ and $f\colon A \to L_1$, respectively.

As it is usual in FCA, the minimal pieces of information are called concepts. In this environment, a *multi-adjoint concept* is a pair $\langle g, f\rangle$ with $g \in L_2^B$, $f \in L_1^A$ satisfying the equalities $g^{\uparrow} = f$ and $f^{\downarrow} = g$. From the consideration of all the concepts associated with the context we can build a complete lattice, called concept lattice, this notion is recalled in the following definition.

Definition 3 The *multi-adjoint concept lattice* associated with a multi-adjoint frame $(L_1, L_2, P, \&_1, \ldots, \&_n)$ and a context (A, B, R, σ) given, is the set

$$\mathcal{M}(A, B, R, \sigma) = \{\langle g, f\rangle \mid g \in L_2^B, f \in L_1^A, g^{\uparrow} = f, f^{\downarrow} = g\}$$

where the ordering is defined by $\langle g_1, f_1\rangle \preceq \langle g_2, f_2\rangle$ if and only if $g_1 \preceq_2 g_2$ (equivalently $f_2 \preceq_1 f_1$).

Henceforth, we write $\mathcal{M}$ instead of $\mathcal{M}(A, B, R, \sigma)$ in order to simplify the notation, as long as there is no possibility of confusion.

The following theorem, presented in [7], shows that the entire concept lattice can be obtained from a specific family of fuzzy subsets of attributes. This special family of attributes as well as the mentioned result are introduced below.

Definition 4 For each $a \in A$, the fuzzy subsets of attributes $\phi_{a,x} \in L_1^A$ defined, for all $x \in L_1$, as

$$\phi_{a,x}(a') = \begin{cases} x & \text{if } a' = a \\ \perp_1 & \text{if } a' \neq a \end{cases}$$

will be called *fuzzy-attributes*, where $\perp_1$ is the minimum element in L_1. The set of all fuzzy-attributes will be denoted as $\Phi_A = \{\phi_{a,x} \mid a \in A, x \in L_1\}$.

Theorem 1 *The set of $\wedge$-irreducible elements of $\mathcal{M}$, $M_F(A)$, coincides with the pairs $\langle \phi_{a,x}^{\downarrow}, \phi_{a,x}^{\downarrow\uparrow}\rangle$ in $\mathcal{M}$, with $a \in A$ and $x \in L_1$, such that*

$$\phi_{a,x}^{\downarrow} \neq \bigwedge\{\phi_{a_i,x_i}^{\downarrow} \mid \phi_{a_i,x_i} \in \Phi_A, \phi_{a,x}^{\downarrow} \prec_2 \phi_{a_i,x_i}^{\downarrow}\}$$

and $\phi_{a,x}^{\downarrow} \neq g_{\top_2}$, where $\top_2$ is the maximum element in L_2 and $g_{\top_2}\colon B \to L_2$ is the fuzzy subset defined as $g_{\top_2}(b) = \top_2$, for all $b \in B$.

The previous result relates the concepts in the concept lattice to the fuzzy-attributes and, therefore, to the attributes. This relation gives rise to the notion of *attribute generating a concept*, which will be crucial in this work (more information about this notion can be found in [7]).

Definition 5 Given a multi-adjoint frame $(L_1, L_2, P, \&_1, \ldots, \&_n)$, a context (A, B, R, σ) associated with the concept lattice $(\mathcal{M}, \preceq)$ and a concept C of it, *the set of attributes generating* C is defined as the set:

$$\text{Atg}(C) = \{a \in A \mid \text{there exists} \quad x \in L_1 \quad \text{such that} \quad \langle \phi_{a,x}^{\downarrow}, \phi_{a,x}^{\downarrow\uparrow} \rangle = C\}$$

Once we have presented all the necessary notions, in the following section we introduce the contributions of this study.

3 Knowledge-Implications in Multi-adjoint Concept Lattices

As it was mentioned in the introduction, in many real situations data contained in a database are redundant. In [4], considering the fuzzy environment of the multi-adjoint concept lattices, it was proven that all the irreducible concepts of a multi-adjoint concept lattice can be obtained from the fuzzy attributes. In particular, it was shown that one concept of the concept lattice can be generated from more than one attribute. This is a very important difference from the classical case and it was studied more deeply in [7], in this work it was presented several possible situations in which a concept $C \in \mathcal{M}$, satisfies that the cardinality of the set $\text{Atg}(C)$ is greater than 1. It is natural to think that this fact gives us information about the knowledge that a specific attribute or a subset of attributes can provide and, consequently, this information can be considered to establish relations among the attributes of our database, depending on the concept generated by them. These relations lead us to reduce the number of attribute, removing the unnecessary attributes.

In this work, these relations are given by means of knowledge implications. The following definition introduces a new notion of implication between subsets of attribute of a multi-adjoint context. The idea that underlies this definition is the following: a subset of attributes implies another one, when the knowledge provided by the subset of attributes in the consequent can be also obtained from the attributes in the antecedent.

Definition 6 Given a multi-adjoint frame $(L_1, L_2, P, \&_1, \ldots, \&_n)$, a multi-adjoint context (A, B, R, σ), the associated concept lattice $(\mathcal{M}, \preceq)$ and two subsets of attributes $A_1, A_2 \subseteq A$. We say that A_1 implies A_2, $A_1 \rightarrow A_2$, if for each $C \in \mathcal{M}$ and $a \in A_2$, such that $a \in Atg(C)$, then there exist $a_1, ..., a_n \in A_1$ and $C_1, ..., C_n \in \mathcal{M}$ satisfying that $a_i \in Atg(C_i)$, for each $1 \leq i \leq n$ and $C = \bigwedge_{1 \leq i \leq n} C_i$.

It should be noted that the previous definition has been developed in terms of infimum, but it can be developed using supremum in an analogous way.

Considering this definition, if we have that $A_1 \rightarrow A_2$, then we can remove the attributes of A_2, since the knowledge obtained from the attributes in A_2 can be deduced from the attributes in A_1.

Note that, when $A_1 \to A_2$ and there exists an attribute $a \in A_2$ with $a \in \text{Atg}(C)$, where $C \in M_F(A)$, since C cannot be obtained from other concepts, then we can find an attribute $a' \in A_1$ satisfying that $a' \in \text{Atg}(C)$.

It is convenient to highlight that the developed implication is a crisp implication between crisp subsets of attributes, within the fuzzy environment of the multi-adjoint concept lattices. These implications do not have associated any truth values and the considered attributes in the implications do not have an assigned truth degree either. This is an important fact to the user, since we must handle imperfect and uncertain information but the final result should be as clear and specific as possible. Therefore, it does not suppose a generalization of other existing attribute implication, rather it is a novel approach, which can be used, e.g. to obtain reducts and minimal reducts in a more efficient way.

It is interesting to analyze if the previously introduced definition of implication satisfies "desirable" properties for its application. Next, five basic properties are presented.

Proposition 1 *Given $A_1, A_2, A_3 \subseteq A$ such that $A_1 \to A_2$. Then the following properties hold:*

1. *If $A_2 \to A_3$ then $A_1 \to A_3$*
2. $A_1 \to A_1$
3. *If $A_1 \to A_3$ then $A_1 \to A_2 \cup A_3$*
4. *If $A_3 \to A_2$ then $A_1 \cup A_3 \to A_2$*
5. *If there exists $a \in A_1$ such that $A_1 \setminus \{a\} \to \{a\}$ then $A_1 \setminus \{a\} \to A_2$*

In the following example, given a multi-adjoint frame and a multi-adjoint context, we illustrate the previously introduced notion and obtain the existing implications among the attributes.

Example 1 Let $(L, \leq, \&_G^*)$ be a multi-adjoint frame, where $\&_G^*$ is the discretization of the Gödel t-norm with respect to $L = \{0, 0.5, 1\}$. In this case, we consider the context (A, B, R, σ), in which $A = \{a_1, a_2, a_3, a_4, a_5\}$, $B = \{b_1, b_2, b_3\}$, $R\colon A \times B \to L$ is displayed in the left side of Fig. 1 and σ is constantly $\&_G^*$. The associated concept lattice $(\mathcal{M}, \preceq)$ has 8 concepts, which is represented in the right side of Fig. 1.

In addition, the set of concepts generated from fuzzy attributes are:

$$C_0 = \langle \phi_{a_5,1.0}^{\downarrow}, \phi_{a_5,1.0}^{\downarrow\uparrow} \rangle$$
$$C_1 = \langle \phi_{a_2,1.0}^{\downarrow}, \phi_{a_2,1.0}^{\downarrow\uparrow} \rangle$$
$$C_3 = \langle \phi_{a_1,0.5}^{\downarrow}, \phi_{a_1,0.5}^{\downarrow\uparrow} \rangle = \langle \phi_{a_1,1.0}^{\downarrow}, \phi_{a_1,1.0}^{\downarrow\uparrow} \rangle = \langle \phi_{a_2,0.5}^{\downarrow}, \phi_{a_2,0.5}^{\downarrow\uparrow} \rangle = \langle \phi_{a_5,0.5}^{\downarrow}, \phi_{a_5,0.5}^{\downarrow\uparrow} \rangle$$
$$C_5 = \langle \phi_{a_3,1.0}^{\downarrow}, \phi_{a_3,1.0}^{\downarrow\uparrow} \rangle$$
$$C_6 = \langle \phi_{a_4,1.0}^{\downarrow}, \phi_{a_4,1.0}^{\downarrow\uparrow} \rangle$$
$$C_7 = \langle \phi_{a_3,0.5}^{\downarrow}, \phi_{a_3,0.5}^{\downarrow\uparrow} \rangle = \langle \phi_{a_4,0.5}^{\downarrow}, \phi_{a_4,0.5}^{\downarrow\uparrow} \rangle = \langle \phi_{a_i,0.0}^{\downarrow}, \phi_{a_i,0.0}^{\downarrow\uparrow} \rangle \quad \text{for all} \quad i \in \{1, 2, 3, 4, 5\}$$

R	b_1	b_2	b_3
a_1	1	1	0
a_2	0.5	1	0
a_3	1	1	0.5
a_4	1	0.5	1
a_5	0.5	0.5	0

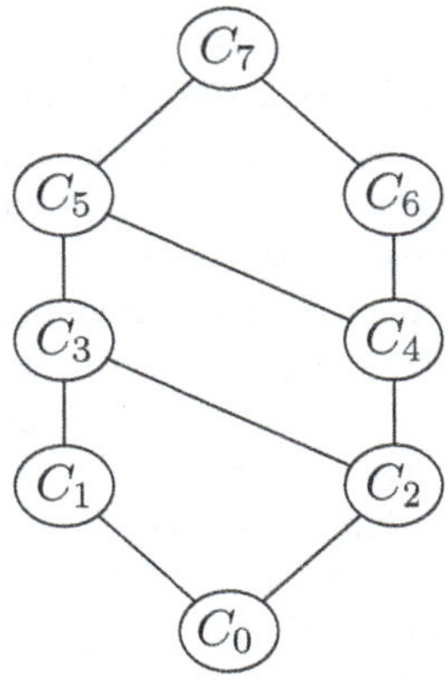

Fig. 1 Relation R (left side) and Hasse diagram of $(\mathcal{M}, \preceq)$ (right side) of Example 1

From the previous list, considering Definition 5 we obtain the set of attributes generating the concepts in $(\mathcal{M}, \preceq)$. In particular, in this example, we obtain that $\text{Atg}(C_0) = \{a_5\}, \text{Atg}(C_1) = \{a_2\}, \text{Atg}(C_3) = \{a_1, a_2, a_5\}, \text{Atg}(C_4) = \varnothing, \text{Atg}(C_5) = \{a_3\}$ and $\text{Atg}(C_6) = \{a_4\}$. Furthermore, according to the concept lattice shown in Fig. 1, we have that $M_F = \{C_1, C_3, C_5, C_6\}$. From all this information, together with the new notion of knowledge implication introduced in Definition 6, we obtain the following set of knowledge implications between attributes:

- $a_5 \to a_1$: since a_1 only generates the concept C_3 which is also generated from a_5.
- $a_2 \to a_1$: since a_2 generates also the concept C_3.
- $\{a_2, a_4\} \to a_5$: since a_5 generates C_3 and C_0. In addition, a_2 and a_4 generate the concepts C_1 and C_6, respectively, and $C_0 = C_1 \wedge C_6$.

The rest of implications among attributes of the context can be obtained from the previous ones by means of the properties given in Proposition 1. For example, the implication $\{a_2, a_4\} \to a_1$ can be deduced applying the transitivity property to the implications $a_5 \to a_1$ and $\{a_2, a_4\} \to a_5$. □

4 Conclusions and Future Works

In this work we have presented a new definition of implication. This implications are built from the perspective of the knowledge obtained from the concept lattice. As a consequence, the considered philosophy to build knowledge implications is different from the usually considered for attribute implication in FCA. Moreover, different properties and an illustrative example have been presented.

As future work, we will analyze in depth the influence of this kind of implications in the reduction of formal contexts. Specifically, we will study the relationship of the set of obtained implications to the attribute classification of the system. In addition, the relationship between the attribute implications usually considered in FCA and the knowledge implications will be studied.

References

1. Baixeries, J., Codocedo, V., Kaytoue, M., Napoli, A.: Characterizing approximate-matching dependencies in formal concept analysis with pattern structures. Discrete Appl. Math. **249**, 18–27 (2018)
2. Bartl, E., Konecny, J.: L-concept analysis with positive and negative attributes. Inf. Sci. **360**, 96–111 (2016)
3. Belohlavek, R., Cordero, P., Enciso, M., Mora, A., Vychodil, V.: Automated prover for attribute dependencies in data with grades. Int. J. Approx. Reason. **70**, 51–67 (2016)
4. Cornejo, M.E., Medina, J., Ramírez-Poussa, E.: Attribute reduction in multi-adjoint concept lattices. Inf. Sci. **294**, 41–56 (2015)
5. Cornejo, M.E., Medina, J., Ramírez-Poussa, E.: Multi-adjoint algebras versus non-commutative residuated structures. Int. J. Approx. Reason. **66**, 119–138 (2015)
6. Cornejo, M.E., Medina, J., Ramírez-Poussa, E.: Adjoint negations, more than residuated negations. Inf. Sci. **345**, 355–371 (2016)
7. Cornejo, M.E., Medina, J., Ramírez-Poussa, E.: Characterizing reducts in multi-adjoint concept lattices. Inf. Sci. **422**, 364–376 (2018)
8. Ganter, B., Wille, R.: Formal Concept Analysis: Mathematical Foundation. Springer (1999)
9. Kuhr, T., Vychodil, V.: Fuzzy logic programming reduced to reasoning with attribute implications. Fuzzy Sets Syst. **262**, 1–20 (2015)
10. Liñeiro-Barea, V., Medina, J., Medina-Bulo, I.: Generating fuzzy attribute rules via fuzzy formal concept analysis. In: Kóczy, L.T., Medina, J. (eds.) Interactions Between Computational Intelligence and Mathematics. Studies in Computational Intelligence, pp. 105–119. Springer, Cham (2018)
11. Medina, J., Ojeda-Aciego, M., Ruiz-Calviño, J.: Formal concept analysis via multi-adjoint concept lattices. Fuzzy Sets Syst. **160**(2), 130–144 (2009)
12. Rodríguez-Jiménez, J.M., Cordero, P., Enciso, M., Mora, A.: Data mining algorithms to compute mixed concepts with negative attributes: an application to breast cancer data analysis. Math. Methods Appl. Sci. **39**(16), 4829–4845 (2016)
13. Rodríguez-Lorenzo, E., Cordero, P., Enciso, M., Mora, A.: Canonical dichotomous direct bases. Inf. Sci. **376**, 39–53 (2017)
14. Vychodil, V.: Computing sets of graded attribute implications with witnessed non-redundancy. Inf. Sci. **351**, 90–100 (2016)

Fuzzy Decision Support Methodology for Sustainable Packaging System Design

Adrienn Buruzs, Kata Vöröskői, Daniel Pup, and László T. Kóczy

Abstract The aim of the present paper is to develop an integrated method that provides assistance to decision makers during packaging system planning, design, operation and evaluation from an environmental perspective. The role of the packaging system is to provide a cover for the handling and communication functions surrounding the product. Single-use and reusable packaging are known based on the time it participates in the goods trade. The purpose of the authors is to develop an evaluation model for the selection of packaging systems from an environmental and sustainability point of view in the supply chain.

Keywords Packaging · Model · Environmental aspects · Decision-making

1 Introduction

1.1 General Introduction and Motivation

The choice of packaging is a key when environmental issues are addressed in logistics. The option of packaging system influences both economic and environmental performance of a supply chain. Current models and methods used to support packaging selection in manufacturing companies rarely consider both of these dimensions. This lack of holistic focus is in line with the general trend which usually focuses on standalone topics of environmental and social issues [1].

A. Buruzs (✉)
Department of Environmental Engineering, Széchenyi István University, Győr, Hungary
e-mail: buruzs@sze.hu

K. Vöröskői
Department of Logistics and Forwarding, Széchenyi István University, Győr, Hungary

D. Pup
Department of Road and Rail Vehicles, Széchenyi István University, Győr, Hungary

L. T. Kóczy
Department of Automation, Széchenyi István University, Győr, Hungary

M. E. Cornejo et al. (eds.), *Computational Intelligence and Mathematics for Tackling Complex Problems 2*, Studies in Computational Intelligence 955,
https://doi.org/10.1007/978-3-030-88817-6_19

According to a recent study [2], the annual growth rate of packaging use will be around 4% by 2019. Successful co-operation between several fields of expertise is needed to create packaging, such as research, design, manufacturing, marketing and advertising, as well as good business spirit, trained staff and some courage. New technologies are emerging, leading to a paradigm shift, which takes place in a very short time. Consumer behavior and decision-making is a very complex issue, influenced by many factors.

The concept of sustainability has gained great importance and attention over the past 15–20 years, a sustainable environment is no longer just a thought, a concept, but a basic requirement, not only in the packaging industry, but in all areas and in all cultures. The same applies to the packaging industry, since packaging is being considered an integral part of the business plan.

There are significant factors that can affect the packaging systems. For instance, the loads can cause a quantitative or qualitative change in the product in the absence of adequate protection. In all cases, the expected impacts, which can have adverse effects from mechanical, climatic, biological and human activities, must be known to develop technically and economically optimal packaging. Preparations for environmental impact and packaging design are supported by laboratory simulation tests. Protecting the environment is a double task: on the one hand, the packaging must ensure that the product does not cause damage in our natural and built environment, and on the other hand, packing has the lowest possible environmental impact throughout its full lifetime. Packaging accompanies and facilitates spatial and temporal processes between production and consumption by making material flow more efficient, faster and more economical as a management unit and providing information flow during logistics operations.

1.2 Structure of the Paper

The structure of the paper is as followings. In the 1.3 paragraph, the literature review is provided. In Chap. 2, an overview on the applied method, namely the Fuzzy Signature Model is given. In Chap. 3, the results of the model's simulation are introduced.

At the end of the paper, the authors concluded that for carrying out the proposed assessment methodology, expert knowledge is needed to ensure the reliability of the results obtained. The future research intentions are summarized in Chap. 5.

1.3 Literature Review

Packaging is now a complex feature. The quality, design, and value of the packaging are close to the value of the product. These functions, however, need to be examined from several angles. Traditionally, packaging has three functions: ensuring the transportability and storage of the goods; preservation of the quality of the goods and the promotion of the goods.

Regarding the use of packaging, it can be single-use (also referred as disposable or one-way) or reusable (also referred as returnable or returning), mainly for consumer and transport packaging. In case of single-use packaging, after use, it becomes waste immediately which means wasting a lot of raw materials, energy and money, and enormous effect on the environment.

A typical example for reusable packaging is the glass bottle which is 50–60 times refillable and can be entirely recycled.

The expectations of different actors in the consumer chain are different. The main interest of the manufacturer is that the product can be transported to the consumer in the most economical way possible without depreciation. The trader wants easy, well-managed packaging. The buyer expects a comfortable, practical package that will fit his own requirements. Packaging must meet all of the major criteria listed below, which is not an easy task:

- protect the product from harmful physical and chemical stress,
- provide a rational management, delivery, sales unit,
- preserve the aroma, smell of the product
- increase product shelf life
- provide communication and ancillary services,
- attracting the attention of the consumer,
- promoting sales,
- being environmentally friendly,
- minimize the cost of loading the product.

The most commonly used packaging materials are in the increasing order of environmental damage: paper, wood, glass, steel sheet, tin-plated steel sheet, aluminum, plastic, chlorine and fluorine-containing plastics (e.g. PVC). From a packaging technology point of view, most of the plastics have an environmental impact, because these are difficult to break down and decompose only over a long period of time.

From the aspect of environmental protection, packaging is considered to be the optimal one that ensures that the product does not become waste or only with the smallest amount of material used. In recent years, waste disposal capacity is getting more and more limited in most countries, and increasing concern has been voiced about the need to curb down on unnecessary waste production caused by single-use packaging [4].

Packaging can be sustainable—irrespectively of its material—(a) if it performs in the same way as competing products, but has a smaller ecological footprint, (b) if its mass is as small as possible and well recyclable. Further, (c) if it contains partially or wholly recycled material, or comes from a renewable raw material source, or is biodegradable or compostable.

The relationship between packaging and the environment today is basically determined by the quantity and rate of packaging waste. The higher the proportion of recycled waste, and the greater the proportion of waste that is being recycled into the materials, the less the burden on the environment is.

A new approach, the Circular Economy strategy comes from the European Union and starts from the realization that our Earth's energy and raw materials are finite, and our planet is scarce in these sources, vulnerable, and our current management of materials is wasteful. In the last one and a half decade, significant price hikes have occurred along with considerable uncertainties in the supply of raw materials. In an economic model called 'linear' by experts, the products are made from fresh raw materials, then sold, used and finally discarded or disposed of as waste. In this process, there are only a few products that have been collected and recycled, and the materials are designed for single use (Fig. 1) [3].

The use of energy sources and raw materials available in finite quantities must be radically changed. In the field of energy, renewable sources should be preferred and materials should be circulated within the limits of reasonableness. The circular economy distinguishes between organic, renewable and extractive mrials. The recirculation of both material flows into the circulation is considered necessary (Fig. 2).

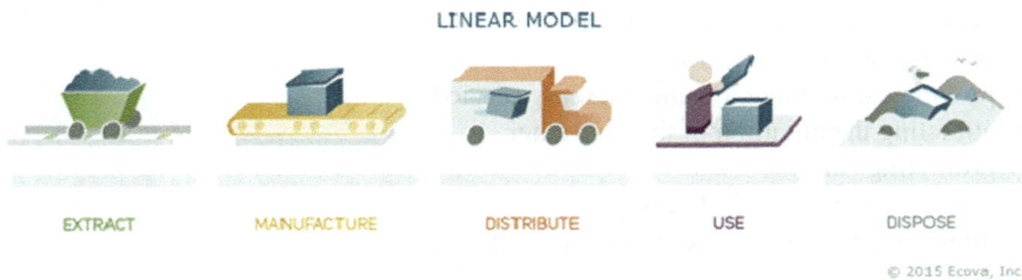

Fig. 1 The linear economic model

Fig. 2 The model of circular economy

Several economic models have been developed to implement the circular economy principle. According to the concept of the first model—especially in the case of durable goods—products should be designed for a long life, so the aim is to produce machines and equipment that will be used for long.

From the above, it can be seen that the circular economy strategy is not a simple waste management issue, but it creates an entirely new economic model, where the key issue is product design. Keeping packaging materials in circulation is a source of resource saving, and needs to improve the efficiency of existing collection systems, and to strengthen awareness-raising work.

1.4 Preliminaries of the Research

In the field of logistics packaging [5], companies have to take decisions on determining the optimal packaging solutions and expenses. The decisions often involve a choice between one-way (disposable) and reusable (returnable) packaging solutions. Even nowadays, in most cases the decisions are made based on traditions and mainly consider the material and investment costs. Although cost is an important factor, it might not be sufficient for finding the optimal solution. Traditional (two-valued) logic is not suitable for modelling this problem, so in the earlier work the authors applied a fuzzy signature approach where three different fuzzy signatures connected by fuzzy rules modelling the packaging decision were suggested, based on logistics expert opinions, in order to support the decision making process of choosing the right packaging system. As a conclusion the authors stated that the improved model with the additional two signatures is even more suitable for decision support than the previous basic model they proposed. However, further research is needed towards refining the model, especially modelling thess environmental issues in more detail.

In this research, in order to model the problem under investigation, a method based on of a combined fuzzy and statistical approach [6] was applied. These latter results raise some new questions, including the idea of the use of undeterministic graphs, thus resulting in Fuzzy Signatures. The method could be applied to other similar multicomponent vague data pools.

Fuzzy Signatures have a special role in data mining. The semantics of the fuzzy concept is relevant issue of data interpretation. Delicate problems of sensitivity and the re-thinking of principal component and factor analysis for fuzzy data are current research issues [7].

2 Methodology Applied

2.1 The Fuzzy Signature Model

Due to the hierarchical ordering of aspects, we found it useful to use the fuzzy signature model. Takahashi et al. [8] further analyzed the notion of fuzzy signature, which is a signature scheme that allows a signature to be generated using 'fuzzy data' (i.e. a noisy string such as a biometric feature) as a signing key, without using any additional user-specific data (such as a helper string in the context of fuzzy extractors). They gave a generic construction of a fuzzy signature scheme from the combination of an ordinary signature scheme with some homomorphic properties regarding keys and signatures, and a new primitive that they call linear sketch, and showed a concrete instantiation based on the Waters signature scheme. A major weakness of their scheme is that fuzzy data is assumed to be distributed uniformly, and another is that it has somewhat large public parameter (proportional to the security parameter), and requires bilinear groups, and either (or both) of these properties could be barriers for implementation and/or practical use.

With the use of the fuzzy signature structure, complex decision model in economic and medical field should be able to be constructed more effectively [9]. Also it is very efficient in other fields of science such as logistics and forwarding.

Each of these signature contains information relevant to the particular data point $\times$ 0: by going higher in the signature structure, less information will be kept. In some operations it is necessary to reduce and aggregate information to become compatible with information obtained from another source (some detail variables missing or simply being locally omitted). Such is when interpolation within a fuzzy signature rule base is done, where the fuzzy signature flanking an observation are not exactly of the same structure. In this case the maximal common sub-tree must be determined and all signature must be reduced to that level in order to be able to interpolate between the corresponding branches or roots in some cases [9].

3 Results

3.1 External Factors

In this paper, the authors focused on the external factors of the complex model introduced in their earlier publication [5]. Based on the above presented method, the authors proposed an aggregation of the leaf-leaf or leaf-node connections.

In the following the tree and sub-tree levels (leaves of the FSig) are described in the group of external factors: weights of the weighted arithmetic means options (w_i) with the chosen aggregations of the (the weighted means) intermediate nodes are listed in Table 1.

Table 1 Ranking and arithmetic means aggregation weights combining the external factors and their respective sub-trees

ID	Features	Ranking	Weights
3	**External factors**		
11231	Cooperation	1	8
11232	Regulations	2	3
11233	Legal	3–4	2
11234	Environmental effects	3–4	2
3.2	**Regulations**		
112321	Environmental	1–4	2
112322	Health	1–4	2
112323	Benefits	1–4	2
112324	Standards	1–4	2
3.4	**Environmental effects**		
112341	Production related	1	10
112342	CO_2 emission	2	8
112343	Pool size	3	7
112344	Effective vehicle utilization	4	5
3.4.1	**Production related**		
1123411	Raw material	1	10
1123412	Energy	2	8

This attribute represents the external conditions, regulations and legal aspects. The degree of cooperation among the participants is a very important aspect and it fundamentally determines the possibility of using returnable packaging systems. Environmental effects cannot always be clearly expressed and considered enough in corporate practice, but they are necessary to be built in the model. These are the following: quantity of raw materials and energy consumed in production, CO_2 emission during return transportation, effective vehicle utilization and pool size (it means the total quantity of returnable packaging devices circulating in the system to ensure the stabile operation) [5].

3.2 *Main Aspects*

In this section the structure of the fuzzy signature modelling our problem on hand will be proposed including the tree graph and the aggregation operations in the intermediate nodes [7]. In the FSig in the intermediate nodes the use of weighted arithmetic mean operations is proposed.

All leaves of the tree assume their values (μ_i) from the interval [0,1]. The values belonging to the intermediate nodes are calculated by respective functions specified to each leaf according to the logistics meaning and role. The relations among the individual descendants on the same level are determined with respective aggregations. When the final value created by the aggregation in the root node (a_0) is close to 0 or it is close to 1, it will present our output decision.

3.3 Aggregation Operators

The variables that influence the decision also affect each other. Because there are many such variables, there is a high degree of uncertainty in the model. In solving such a problem, where human factors play a role, it is imperative to arrange the aspects into hierarchical groups. This is why the authors found it useful to use the fuzzy signature model.

During the application it became clear that further aspects had to be taken into account. The three levels resulted in another decision. Based on the modeling of the problem, it was found that in principle it would be possible to use a simple fuzzy signature model, but the decision groups are built on each other to such an extent that a sequence of decisions is needed. In this sense, the authors propose a novel model consisting of multiple fuzzy signatures of hierarchical cascade-like subordination (Fig. 3.).

These signatures have fixed numbers and multiple decisions are required. Aggregations were determined by knowing the expert domain. Aggregations were used to evaluate the first signature and, if the fuzzy recommendation membership value is below 0.5, one-way packaging is recommended. If fuzzy has a recommendation membership value of 0.5 or greater, another signature will still follow.

If $S_1 \geq 0.5$ then single-use, if $S_2 \geq 0.5$ then either, if $S_3 \geq 0.5$ then returnable packaging is recommended.

Based on the opinion of a panel of logistics experts is was decided that all aggregations are of the weighted arithmetic mean type, because the components of the individual characteristics and features are comparably importance, which may be expressed by weights of the same order of magnitude [7]. When the final value created by the aggregation in the root (a_0) is close to 0, it should rather be one-way, if the result of a_1 is close to 1, the packaging should rather be returnable.

Based on the relationships assumed by environmental experts and because the components of the individual characteristics and features are comparably importance, the following aggregations have been suggested:

- Aggregation between leaf and leaf

  μ_{211} (distance) and μ_{342} (CO_2 emission)
  μ_{212} (volume) and μ_{3411} (raw material)
  μ_{232} (FTL/LTL) and μ_{344} (effective vehicle utilization)

- Aggregation between leaf and non-leaf

  a_{341} (production related) and μ_{212} (volume)
  a_{32} (regulation) and μ_{33} (legal)

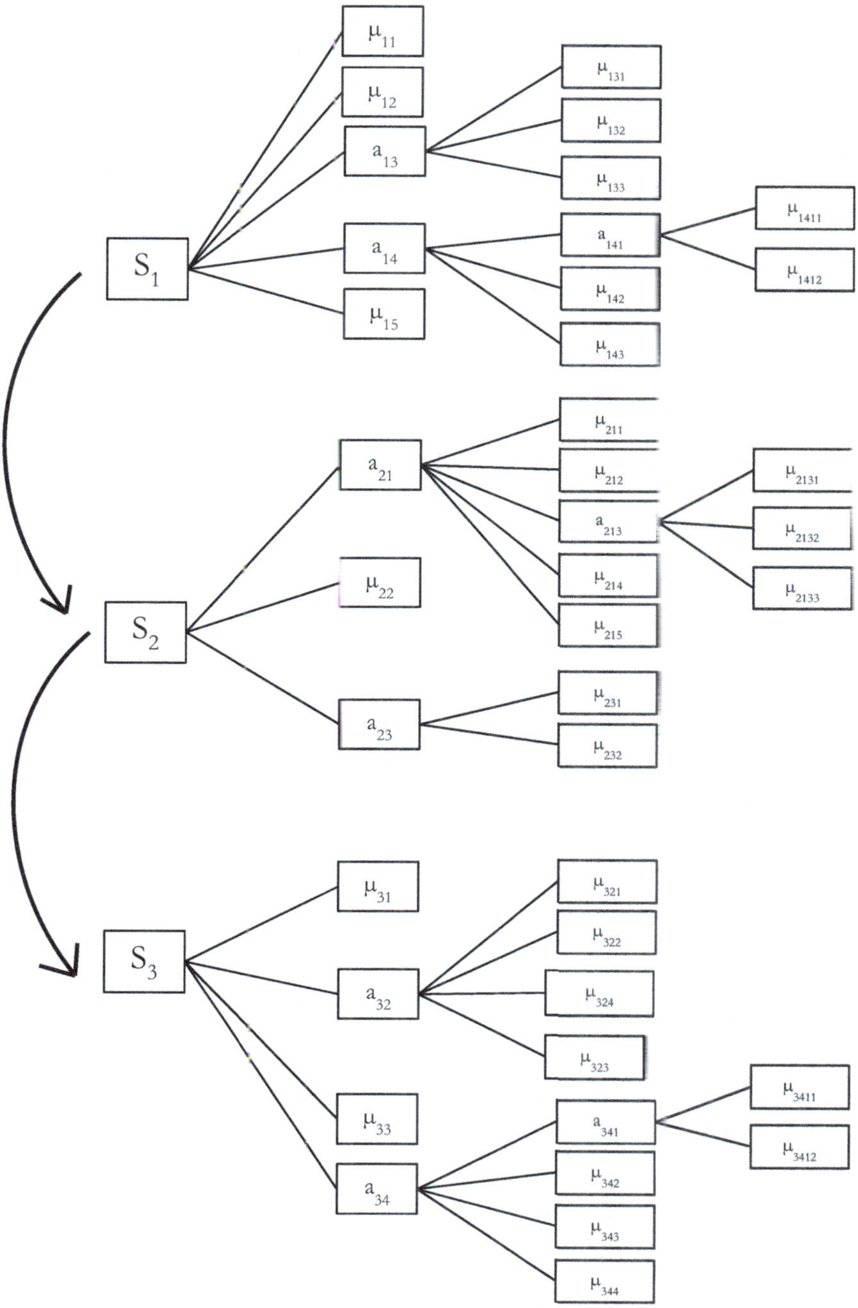

Fig. 3 Multiple fuzzy signatures of hierarchical cascade-like subordination

a_{34} (environmental effects) and μ_{321} (environmental)

The result of the first FSig was calculated by all aggregations according to Table 1 is $a_I = 0.519534$. If we recalculate it with the additional signatures, the result in the improved model is $a_0 = 0.519445$.

According to the aspects considered in the model under investigation, the packaging system should rather be returnable.

4 Conclusions and Future Research

The intermeshing of disciplines from the natural sciences, social sciences, engineering and management is essential to address this complex problem. The study presented here possesses also this feature for future applications.

The authors' purpose is to continue the investigation to understand the deeper context of the circular economy and try to develop a refined model.

Everybody in the world should be aware of packaging in order to develop a sustainable consumer society. Efficiency is always the responsibility of the manufacturer, and since the packaging has a mission, the packaging materials should be used properly. In this area, the overarching task is to ensure that all packaging and packaging we use reach its destination.

The way forward is the selection and application of the least burdensome packaging materials for the environment. The selection of the best packaging materials can be done by life cycle assessments (LCA). The LCA extends from the extraction of raw materials to the final waste remaining after the use of the product. The following aspects can be used to judge packaging materials from the environmental side:

- the packaging material production uses the natural resources (e.g. energy sources, water, forest) as little as possible;
- the production of packaging materials and equipment should not cause environmental damage
- follow the most cost-effective solution when designing the packaging and choosing the packaging material;
- reuse the used packaging material or recycle it
- if recycling is not feasible, the packaging material can be incinerated or composted without harmful emissions.

On the top of that, packaging is often responsible for a considerable fraction of the energy and material requirements associated to the life cycle of a product, and is thus also detrimental to the latter's overall sustainability, both in terms of resource depletion and of the associated emissions [4].

Acknowledgements The research presented in this paper was carried out as part of the "Dynamics and Control of Autonomous Vehicles meeting the Synergy Demands of Automated Transport Systems (EFOP-3.6.2-16-2017-00016)" project in the framework of the New Széchenyi Plan. The completion of this project is funded by the European Union and co-financed by the European Social Fund.

References

1. Pålsson, H., Finnsgård, C., Wänström, C.: Selection of packaging systems in supply chains from a sustainability perspective: the case of Volvo. Packag. Technol. Sci. **26**(5), 289–310 (2013)
2. Schneider, T.L.: Packaging from a Global Approach – Trends, Challenges. Hungarian Packaging Yearbook (2016–2017)
3. Goellner, K.N., Sparrow, E.: An environmental impact comparison of single-use and reusable thermally controlled shipping containers. Int. J. Life Cycle Assess. **19**(3), 611–619 (2014)
4. Raugeil, M., Fullana-i-Palmer, M., Puig, R., Torres, A.: Single-use vs. reusable transport packaging: a comparative life cycle assessment. Packag. Technol. Sci. **22**(8), 443–450 (2009). http://hdl.handle.net/2117/6533, ISSN0894–3214
5. Vöröskői, K., Fogarasi, G., Buruzs, A., Földesi, P., Kóczy, T.L.: Hierarchical fuzzy decision support methodology for packaging system design. In: Kulczycki, P., Kacprzyk, J., Kóczy, L.T., Mesiar, R., Wisniewski, R. (eds.) Information Technology, Systems Research, and Computational Physics, pp. 85–96. Springer, Cham, Svájc (2020). Paper: Chapter 7, 12 p.
6. Kóczy, L.T., Purvinis, O., Susnienė, D.: A combined fuzzy and least squares method approach for the evaluation of management questionnaires. In: Šostak, A., Ramírez-Poussa, E., Medina-Moreno, J., Kóczy, L.T (eds.) Computational Intelligence and Mathematics for Tackling Complex Problems, pp. 157–165. Springer (2020). Paper: Chapter 20, 9 p.
7. Vamos, T., Koczy L.T., Biro, G.: Fuzzy signatures in data mining. In Proceedings Joint 9th IFSA World Congress and 20th NAFIPS International Conference (Cat. No. 01TH8569) (2001)
8. Matsuda, T., Takahashi, K., Murakami, T., Hanaoka, G.: Fuzzy Signatures: Relaxing Requirements and a New Construction, pp. 97–116 (2016). https://doi.org/10.1007/978-3-319-39555-5_6
9. Wong, K., Gedeon, T., Chong, A., Kóczy, L.T.: Fuzzy signatures for complex decision model Aust. J. Intell. Inf. Process. Syst. **8**, 193–199 (2005)

Some Relationships Between the Notions of f-Inclusion and f-Contradiction

Nicolás Madrid and Manuel Ojeda-Aciego

Abstract In this paper we analyse the relationships between the notions of f-inclusion and f-weak-contradiction. In particular, we present some theoretical results that relate both notions by means of negation operators (used to define complements of fuzzy sets) and Galois connections.

Keywords Galois connections · Inclusion measure · Contradiction measure · Fuzzy sets

1 Introduction

The notions of f-inclusion [12, 13] and f-weak-contradiction [3] were introduced, respectively, to model inclusion and contradiction between fuzzy sets. The distinctive feature of these approaches with respect to others which can be found in the literature [4–6] is that they use mappings, instead of using real numbers, in order to represent the inclusion and contradiction between two given fuzzy sets. In other words, mappings are considered as a degree either of inclusion or contradiction. Interestingly enough, many the of axioms required for measures of inclusion based on real numbers [9, 14, 16, 17], suitably changing real values by mappings [11], also hold in the framework of f-inclusion which, in fact, can be considered as a kind of Sinha-Dougherty inclusion [14].

The notions of f-inclusion and f-contradiction are defined similarly: they both are crisp relations which determine a restriction, via a mapping, in terms of an

Partially supported by the Ministry of Science, Innovation, and Universities (MCIU), the State Agency of Research (AEI) and the European Social Fund (FEDER) through the research project PGC2018-095869-B-I00 (MCIU/AEI/FEDER, UE).

N. Madrid (✉) · M. Ojeda-Aciego
Departamento de Matemática Aplicada, Universidad de Málaga, Málaga, Spain
e-mail: nicolas.madrid@uma.es

M. Ojeda-Aciego
e-mail: aciego@uma.es

M. E. Cornejo et al. (eds.), *Computational Intelligence and Mathematics for Tackling Complex Problems 2*, Studies in Computational Intelligence 955,
https://doi.org/10.1007/978-3-030-88817-6_20

inequality. Such a similarity suggests a strong relationship between both notions, which is studied in this paper. Specifically, we show that given two fuzzy sets, their indexes of f-inclusion and f-contradiction can be related to each other by using negation operators and Galois connections [7].

The paper is structured as follows: in Sect. 2 we recall some basic notions of fuzzy set theory, the notion of f-inclusion and the notion of f-weak-contradiction. Then, in Sect. 3 we show the main contribution of the paper. Finally, in Sect. 4 we present conclusions and prospects of future works.

2 Preliminaries

2.1 Basic Notions About Fuzzy Sets

Let us recall that a fuzzy set A is defined on a referential universe $\mathcal{U}$ (usually omitted) by means of its membership function $A: \mathcal{U} \to [0, 1]$. The standard operations union and intersection between sets can be extended to operations between fuzzy sets as follows: given two fuzzy sets A and B, we define the fuzzy sets $A \cup B$ and $A \cup B$ as $A \cap B(x) = \max\{A(u), B(u)\}$ and $A \cap B(u) = \min\{A(u), B(u)\}$ for all $u \in \mathcal{U}$, respectively. To define the complement of a fuzzy set we need to consider negation operators. Let us recall that a negation operator is a decreasing mapping $n: [0, 1] \to [0, 1]$ such that $n(0) = 1$ and $n(1) = 0$. Given a fixed negation operator, n, the complement of a fuzzy set A is defined as $A^c(u) = n(A(u))$ for all $u \in \mathcal{U}$. We say that a negation n is involutive if $n^2(x) = n(n(x)) = x$ for all $x \in [0, 1]$. Note that in the case of considering an involutive negation for the definition of the complement, we have the double complement law; i.e., $(A^c)^c = A$.

We will also need the notion of (isotone or antitone) Galois connection. The formal definitions are given below:

Definition 1 A pair (f, g) of mappings $f, g: [0, 1] \to [0, 1]$ forms an *isotone Galois connection* if and only if the equivalence below holds for all $x, y \in [0, 1]$:

$$f(x) \leq y \iff x \leq g(y) \tag{1}$$

Dually, A pair (f, g) of mappings $f, g: [0, 1] \to [0, 1]$ forms an *antitone Galois connection* if and only if the equivalence below holds for all $x, y \in [0, 1]$:

$$y \leq f(x) \iff x \leq g(y) \tag{2}$$

2.2 φ-Indexes of Inclusion

The main difference of our approach with respect to other approaches dealing with measures of inclusion between fuzzy sets [9, 14, 16, 17], is that the grades used to express the measure are no longer real values, but certain mappings from [0, 1] to [0, 1]. These mappings should satisfy the properties, namely deflation and monotonicity, it one to be considered degrees of inclusion.

Definition 2 The set of φ-indexes of inclusion (denoted by Ω) is the set of mappings $f\colon [0,1] \to [0,1]$ satisfying the following properties for all $x, y \in L$:

- $f(x) \leq x$;
- if $x \leq y$ then $f(x) \leq f(y)$

The definition of f-inclusion [13] is given as follows.

Definition 3 Let A and B be two fuzzy sets and consider $f \in \Omega$. We say that A *is f-included in* B (denoted by $A \subseteq_f B$) if and only if the inequality $f(A(u)) \leq B(u)$ holds for all $u \in \mathcal{U}$.

It is worth remarking that, despite the fact that f-inclusion is a crisp relation for all f, we can define a fuzzy-like inclusion relation by considering the whole set of f-inclusion relations.

Roughly speaking, note that we can define a lattice structure in Ω by considering the usual pointwise ordering between mappings and, moreover, the greater the mapping $f \in \Omega$ the stronger the restriction by the f-inclusion; since given $A \subseteq_f B$ then, the value $f(A(u))$ determines a lower bound of the possible values of $B(u)$, for all $u \in \mathcal{U}$. In this way, the degree of inclusion of fuzzy set A into fuzzy set B can be defined by choosing the greatest $f \in \Omega$ such that A is f-included in B; for more details we refer the reader to [12]. We denote by id and $\bot$ the greatest and lowest mapping in Ω; i.e., the strongest and the weakest degrees of f-inclusion.

Below we summarise some properties of the notion of f-inclusion which motivate the consideration of mappings in Ω as proper degrees of inclusion [12].

Theorem 1 *Let A, B and C be three fuzzy sets and let $f, g \in \Omega$. Then,*

1. *$A \subseteq_f A$ for all $f \in \Omega$.*
2. *$A \subseteq_f B$ and $B \subseteq_g C$ implies $A \subseteq_{g \circ f} C$*
3. *$A \subseteq_\bot B$;*
4. *$A \subseteq_{id} B \iff A(u) \leq B(u)$ for all $u \in \mathcal{U} \iff A \subseteq_f B$ for all $f \in \Omega$.*
5. *If $f \leq g$ and $A \subseteq_g B$, then $A \subseteq_f B$;*
6. *$A \subseteq_f B$ implies $f = \bot$ if and only if there is a nonempty set of elements in the universe $\{u_i\}_{i \in \mathbb{I}} \subseteq \mathcal{U}$ such that $A(u_i) = 1$ for all $i \in \mathbb{I}$ and $\bigwedge_{i \in \mathbb{I}} B(u_i) = 0$.*
7. *If $\mathcal{U}$ is finite, then there exists $u \in \mathcal{U}$ such that $A(u) = 1$ and $B(u) = 0$ if and only if $A \subseteq_f B$ implies $f = \bot$.*
8. *if A and B are two crisp sets, if $A \not\subseteq_\bot B$ then $A \subseteq_{id} B$.*

2.3 φ-Indexes of Contradiction

As in the case of f-inclusion, the idea with the functional indexes of contradiction is to measure contradiction no longer with real values, but with certain mappings from $[0, 1]$ to $[0, 1]$. Those mapping used as degrees of contradiction should satisfy some properties that resemble to negation operators.

Definition 4 The set of φ-indexes of contradiction (denoted by $\overline{\Omega}$) is the set of mappings $f : [0, 1] \to [0, 1]$ satisfying the following properties for all $x, y \in L$:

- $f(0) = 1$;
- if $x \leq y$ then $f(y) \leq f(x)$

The notion of f-weak-contradiction [2, 3] is a slight generalization of the notion of N-contradiction in [15] given by Trillas et al.

Definition 5 Let A and B be two fuzzy sets defined over a nonempty universe $\mathcal{U}$ and let $f \in \overline{\Omega}$. We say that A *is* f*-weak-contradictory w.r.t.* B if and only if $A(x) \leq f(B(x))$ holds for all $x \in \mathcal{U}$.

As in the case of the f-inclusion, the f-weak-contradiction defines crisp relations that pretends to determine degrees (in this case of contradiction) between fuzzy sets. Note that $\overline{\Omega}$ has a structure of complete lattice with the usual pointwise ordering between mappings and that different mappings in $\overline{\Omega}$ determine different kinds of f-weak-contradiction relations, i.e. a stronger or weaker restriction. Specifically, fixed $f \in \overline{\Omega}$ and a value of $B(x)$ (for some $x \in \mathcal{U}$), the f-weak-contradiction determines an upper bound on the value of $A(x)$.

Note that, as f is antitonic, the greater the value of $B(x)$, the smaller the upper bound, and then also the smaller the value of $A(x)$. Therefore, given two fuzzy sets A and B, a degree of contradiction between A and B can be defined by choosing the least $f \in \overline{\Omega}$ such that A is f-weak-contradictory w.r.t. B. We denote by $f^{\top}$ and $f_{\perp}$ the greatest and lowest mapping in $\overline{\Omega}$; i.e., the weakest and the strongest degrees of f-weak-contradiction; for more details we refer the reader to [3].

Below we summarise some properties of the notion of f-weak-contradiction which motivates the consideration of $\overline{\Omega}$ as a proper set of degrees of contradiction.

Theorem 2 *Let A, B and C be three fuzzy sets and let $f, g \in \overline{\Omega}$. Then,*

1. *A is $f^{\top}$-weak-contradictory w.r.t. B.*
2. *If A is $f^{\perp}$-weak-contradictory w.r.t. B then, A is f-weak-contradictory w.r.t. B.*
3. *If $f \leq g$ and A is f-weak-contradictory w.r.t. B then, A is g-weak-contradictory w.r.t. B as well.*
4. *A is $f_{\perp}$-weak-contradictory w.r.t. B if and only if $B(x) > 0$ implies $A(x) = 0$ for all $x \in \mathcal{U}$.*
5. *If $A \leq B$ and C is f-weak-contradictory w.r.t. B then, C is f-weak-contradictory w.r.t. A;*

6. *If $A \leq C$ and C is f-weak-contradictory w.r.t. B then, A is f-weak-contradictory w.r.t. B;*
7. *$f^{\top}$-weak-contradiction is the only weak-contradiction of A w.r.t. B if and only if there exists a sequence $\{x_i\}_{i\in\mathbb{N}} \subseteq \mathcal{U}$ such that $B(x_i) = 1$ for all x_i and $\lim A(x_i) = 1$.*
8. *If $\mathcal{U}$ is a finite universe, $f^{\top}$-weak-contradiction is the only weak-contradiction of A w.r.t. B if and only if there exists $x \in \mathcal{U}$ such that $A(x) = B(x) = 1$.*
9. *If (f, g) is an antitone Galois connection then, A is f-weak-contradictory w.r.t. B if and only if B is g-weak-contradictory w.r.t. A.*

3 Relationships Between f-Contradiction and f-Inclusion

3.1 From f-Inclusion to f-Contradiction

We assume here an index of f-inclusion between two fuzzy sets and infer a corresponding index of f-weak-contradiction between fuzzy sets as well. Hereafter, we will consider a fixed involutive negation operator n so that the complement A^c of a fuzzy set A always denotes the complement w.r.t. n.

Proposition 1 *Let $f \in \Omega$ and let A and B be two fuzzy sets such that A is f-included in B. Then, B^c is $n \circ f$-weak-contradictory w.r.t. A.*

Proposition 2 *Let $f \in \Omega$ such that (f, g) forms an isotone Galois connection and let A and B be two fuzzy sets such that A is f-included in B. Then, A is $g \circ n$-weak-contradictory w.r.t. B^c.*

3.2 From f-Contradiction to f-Inclusion

Let us study now the converse relation to that given in the previous section, that is, given an f-weak-contradiction between two fuzzy sets, determine a corresponding f-inclusion relationship. The main obstacle here is to relate indexes $f \in \overline{\Omega}$ of f-contradiction with indexes in Ω of f-inclusion. In the first result we avoid such an obstacle by requiring that $n(x) \leq f(x)$.

Proposition 3 *Let $f \in \overline{\Omega}$ such that $n(x) \leq f(x)$ and let A and B be two fuzzy sets such that A is f-weak-contradictory w.r.t. B. Then, B is $n \circ f$-included in A^c.*

In order to obtain a second result relating f-weak-contradiction with f-inclusion, we need to consider Galois connections.

Proposition 4 *Let $f \in \overline{\Omega}$ such that $n(x) \leq f(x)$ and $(n \circ f, g \circ n)$ forms an antitone Galois connection. Let A and B be two fuzzy sets such that A is f-weak-contradictory with B. Then, A is $n \circ g$-included in B^c.*

Summarizing, and collecting a result from [12], we have the following theorem.

Theorem 3 *Let $f \in \Omega$ such that (f, g) forms an isotone Galois connection and let A and B be two fuzzy sets. Then the following items are equivalent*

1. *A is f-included in B.*
2. *A is $(g \circ n)$-weak-contradictory with B^c.*
3. *B^c is $(n \circ f)$-weak-contradictory with A.*
4. *B^c is $(n \circ g \circ n)$-included in A^c.*

4 Conclusion and Future Work

We have presented some relationships between the notions of f-inclusion and f-weak-contradiction. Specifically, we have shown that both notions can be related by means of complement of fuzzy sets (i.e., an involutive negation operator) and isotone Galois connections.

The consideration of indexes of inclusion given by mappings (e.g., by isotone Galois connections) may be of great interest in techniques where Galois connections appear naturally (as Mathematical Morphology [8, 10] or Formal Concept Analysis [1]) since composition is a natural operator that, in those contexts, can be used to establish relationships between similar entities. In addition, relationships between f-inclusion and f-contradiction may open a door for studying further relationships between mathematical morphology and formal concept analysis by means of f-weak-contradiction indexes as well.

References

1. Antoni, L., Cabrera, I.P., Krajči, S., Krídlo, O., Ojeda-Aciego, M.: The Chu construction and generalized formal concept analysis. Int. J. Gen. Syst. **46**, 458–474 (2017)
2. Bustince, H., Madrid, N., Ojeda-Aciego, M.: A measure of contradiction based on the notion of n-weak-contradiction. In: IEEE International Conference on Fuzzy Systems (FUZZ-IEEE'13), pp. 1–6 (2013)
3. Bustince, H., Madrid, N., Ojeda-Aciego, M.: The notion of weak-contradiction: definition and measures. IEEE Trans. Fuzzy Syst. **23**, 1057–1069 (2015)
4. Bustince, H., Mohedano, V., Barrenechea, E., Pagola, M.: Definition and construction of fuzzy DI-subsethood measures. Inf. Sci. **176**(21), 3190–3231 (2006)
5. De Baets, B., Meyer, H., Naessens, H.: A class of rational cardinality-based similarity measures. J. Comput. Appl. Math. **132**(1), 51–69 (2001)
6. De Baets, B., Meyer, H., Naessens, H.: On rational cardinality-based inclusion measures. Fuzzy Sets Syst. **128**, 169–183 (2002)
7. García-Pardo, F., Cabrera, I., Cordero, P., Ojeda-Aciego, M.: On Galois connections and soft computing. Lect. Notes Comput. Sci. **7903**, 224–235 (2013)
8. Heijmans, H., Ronse, C.: The algebraic basis of mathematical morphology I. Dilations and erosions. Comput. Vis. Graph. Image Process. **50**(3), 245–295 (1990)

9. Kitainik, L.M.: Fuzzy inclusions and fuzzy dichotomous decision procedures. In: Kacprzyk, J., Orlovski, S.A. (eds.) Optimization Models Using Fuzzy Sets and Possibility Theory, pp. 154–170. Springer, Netherlands, Dordrecht (1987)
10. Madrid, N., Medina, J., Ojeda-Aciego, M., Perfilieva, I.: L-fuzzy relational mathematical morphology based on adjoint triples. Inf. Sci. **474**, 75–89 (2019)
11. Madrid, N., Ojeda-Aciego, M.: A view of f-indexes of inclusion under different axiomatic definitions of fuzzy inclusion. Lect. Notes Artif. Intell. Sci. **10564**, 307–318 (2017)
12. Madrid, N., Ojeda-Aciego, M.: Functional degrees of inclusion and similarity between L-fuzzy sets. Fuzzy Sets Syst. (2019). To appear
13. Madrid, N., Ojeda-Aciego, M., Perfilieva, I.: f-inclusion indexes between fuzzy sets. In: Proceedings of IFSA-EUSFLAT (2015)
14. Sinha, D., Dougherty, E.R.: Fuzzification of set inclusion: theory and applications. Fuzzy Sets Syst. **55**, 15–42 (1993)
15. Trillas, E., Alsina, C., Jacas, J.: On contradiction in fuzzy logic. Soft Comput. **3**(4), 197–199 (1999)
16. Xie, B., Han, L.-W., Mi, J.-S.: Measuring inclusion, similarity and compatibility between Atanassov's intuitionistic fuzzy sets. Int. J. Uncertain. Fuzziness Knowl. Based Syst. **20**(02), 261–280 (2012)
17. Young, V.R.: Fuzzy subsethood. Fuzzy Sets Syst. **77**, 371–384 (1995)

Decomposition Integrals for Interval-Valued Functions

Adam Šeliga

Abstract In this contribution we define a new integral for interval-valued functions. This integral is an extension of a decomposition integral for real-valued functions. We present two different extensions of this integral. The first one is based on the integral of Aumann that is an extension of Riemann integral for set-valued functions. The second one is based on endpoint-wise interval operations. Interestingly, both of these at a first sight different approaches lead to the same integral. An illustrative example is added, too.

Keywords Decomposition integrals · Interval-valued functions · Set integrals

1 Introduction

It is well-known fact that the theory of integration plays an important role in mathematics. The theory starts with Riemann integral and mostly deals with integrating single-valued functions.

An easy and inventive idea of creating an envelope of all possible values of Aumann yields an integral for set-valued functions [1]. This, now so-called Aumann integral, is an extension of Riemann integral for set-valued functions. This idea was later applied to other integrals mostly for interval-valued functions generating a large class of integrals of Aumann type. To note few, the Henstock-Kurzweil-Aumann integral [12], the Lebesgue-Aumann integral [7], the Sugeno-Aumann integral [14], the Choquet-Aumann integral [2, 5], or, in a setting of Banach space functions, the Gould-Aumann integral [9].

This work was supported by the Slovak Research and Development Agency under contracts no. APVV-17-0066 and no. APVV-18-0052. Also the support of the grant VEGA 1/0006/19 is kindly announced.

A. Šeliga (✉)
Faculty of Civil Engineering, Slovak University of Technology, Radlinského 11, 810 05 Bratislava, Slovakia
e-mail: adam.seliga@stuba.sk

M. E. Cornejo et al. (eds.), *Computational Intelligence and Mathematics for Tackling Complex Problems 2*, Studies in Computational Intelligence 955,
https://doi.org/10.1007/978-3-030-88817-6_21

Decomposition integrals [4, 8] are nowadays a flourishing topic in research. This is due to the fact that decomposition integrals are a framework that include many different integrals, e.g., the Choquet integral [3], the Shilkret integral [11], the PAN integral [13], the concave integral [6], and due to the applicability of such integrals in real-work applications, e.g., decision making, production optimization, logistics, economy.

This paper is organized as follows. In the second section we examine preliminaries needed for latter. In Sect. 3 we define an Aumann-like decomposition integral based on the integral of Aumann. In Sect. 4 we present a direct modification of the definition of a decomposition integral using interval operations. Surprisingly, these two approaches are equivalent. The last section, Sect. 5, are concluding remarks.

2 Preliminaries

Throughout this paper we will consider a fixed finite space X. Without loss of generality we may assume that $X = \{1, 2, \dots, n\} \subsetneq \mathbb{N}$. A real-valued function, or simply a function, is any mapping $f\colon X \to [0, \infty[$. A class of all such functions will be denoted by $\mathcal{F}$. An interval-valued function is any mapping $v\colon X \to L([0, \infty[)$ where $L([0, \infty[)$ denotes the class of all closed intervals that are a subset of $[0, \infty[$. The class of all interval-valued functions will be denoted by $\mathcal{V}$. In the following, by "function" we always mean a "real-valued function."

A capacity $\mu\colon 2^X \to [0, \infty[$ is any set function on 2^X that is grounded, i.e., $\mu(\emptyset) = 0$, and monotone with respect to the set inclusion, i.e., $A \subseteq B \subseteq X$ implies $\mu(A) \leq \mu(B)$. The class of all capacities is denoted by $\mathcal{M}$.

A collection, mostly denoted by $\mathcal{D}$, is any non-empty subset of 2^X. A decomposition system, mostly denoted by $\mathcal{H}$, is a class consisting of at least one collection, i.e., non-empty subset of $2^{2^X \setminus \{\emptyset\}}$.

Let $[a, b]$ and $[c, d]$ be two closed intervals, i.e., $[a, b], [c, d] \in L([0, \infty[)$. We say that $[a, b]$ is less or equal to $[c, d]$, writing $[a, b] \leq [c, d]$, if and only if $a \leq c$ and $b \leq d$. The sum and the supremum of these intervals is defined endpoint-wise, i.e.,

$$[a, b] + [c, d] = [a + c, b + d],$$
$$[a, b] \vee [c, d] = [a \vee c, b \vee d],$$

and scalar multiplication by non-negative constant α is defined by

$$\alpha[a, b] = [\alpha a, \alpha b].$$

For every interval-valued function $v \in \mathcal{V}$ there are two functions $a_v, b_v \in \mathcal{F}$ such that $v(x) = [a_v(x), b_v(x)]$ for all $x \in X$, i.e., a_v is the left endpoint of v, and, similarly, b_v is the right endpoint of v. A function $f \in \mathcal{F}$ is a selector of $v \in \mathcal{V}$ if and only if $f(x) \in v(x)$ for all $x \in X$. If f is a selector of v we will write $f \in v$.

A decomposition integral [4, 8] with respect to a decomposition system $\mathcal{H}$ is an operator $I_{\mathcal{H}}\colon \mathcal{F} \times \mathcal{M} \to [0, \infty[$ such that

$$I_{\mathcal{H}}(f, \mu) = \bigvee_{\mathcal{D} \in \mathcal{H}} \mathcal{I}_{\mathcal{D}}(f, \mu),$$

where $\mathcal{I}_{\mathcal{D}}$ is a collection integral [10] with respect to a collection $\mathcal{D}$ given by

$$\mathcal{I}_{\mathcal{D}}(f, \mu) = \bigvee \left\{ \sum_{A \in \mathcal{D}} \alpha_A \mu(A) \colon \sum_{A \in \mathcal{D}} \alpha_A 1_A \le f, \alpha_A \ge 0 \right\},$$

where 1_A denotes the characteristic function of the set A, i.e.,

$$1_A \colon X \to \{0, 1\} \colon x \mapsto \begin{cases} 1, & \text{if } x \in A, \\ 0, & \text{otherwise.} \end{cases}$$

3 Aumann-Like Decomposition Integral

In this section we will present an Aumann-like decomposition integral and we will prove some of its properties. Note that, if $v\colon [a, b] \to L([0, \infty[)$ is an interval-valued function, then the Aumann integral [1] of v is given by

$$\text{Aum}_{[a,b]}(v) = \left\{ \text{R}_{[a,b]}(f) \colon f \in v \right\},$$

where $\text{R}_{[a,b]}(f)$ is the Riemann integral of f on $[a, b]$. For interval-valued functions whose endpoints are Riemann-integrable, we have

$$\text{Aum}_{[a,b]}(v) = \left[\text{R}_{[a,b]}(a_v), \text{R}_{[a,b]}(b_v) \right].$$

Analogous result will be proved for Aumann-like decomposition integral.

Definition 1 Let $\mathcal{H}$ be a decomposition system. An Aumann-like decomposition integral with respect to the decomposition system $\mathcal{H}$ is an operator

$$\hat{I}_{\mathcal{H}} \colon \mathcal{V} \times \mathcal{M} \to 2^{[0,\infty[} \colon (v, \mu) \mapsto \{ I_{\mathcal{H}}(f, \mu) \colon f \in v \}.$$

It is easy to see that $a_v, b_v \in v$ and thus we obtain trivially the following proposition.

Proposition 1 *Let $\mathcal{H}$ be a decomposition system, let $v \in \mathcal{V}$ be an interval-valued function, and let $\mu \in \mathcal{M}$ be a capacity. Then*

$$\{I_{\mathcal{H}}(a_v, \mu), I_{\mathcal{H}}(b_v, \mu)\} \subseteq \hat{I}_{\mathcal{H}}(v, \mu).$$

In the next proposition we will prove that not only the endpoints of v do belong to $\hat{I}_{\mathcal{H}}(v, \mu)$ but also any values in-between them.

Proposition 2 *Let $\mathcal{H}$ be a decomposition system, let $f, g \in \mathcal{F}$ be two functions such that $f \leq g$, and let $\mu \in \mathcal{M}$ be a capacity. Then there is a function $h_\gamma \in \mathcal{F}$ for every $\gamma \in [I_{\mathcal{H}}(f, \mu), I_{\mathcal{H}}(g, \mu)]$ such that $f \leq h_\gamma \leq g$ and $I_{\mathcal{H}}(h_\gamma, \mu) = \gamma$.*

Proof Let us denote $\alpha = I_{\mathcal{H}}(f, \mu)$ and $\beta = I_{\mathcal{H}}(g, \mu)$. From monotonicity we have that $\alpha \leq \beta$ and thus $[\alpha, \beta]$ is indeed an interval. Now let $\gamma \in [\alpha, \beta]$ be any real number between α and β. Let us construct two sequences of functions $\{l_n\}_{n\in\mathbb{N}}$ and $\{r_n\}_{n\in\mathbb{N}}$ defined recursively by $l_1 = f, r_1 = g$,

$$l_{n+1} = \begin{cases} (l_n + r_n)/2, & \text{if } I_{\mathcal{H}}((l_n + r_n)/2, \mu) \leq \gamma, \\ l_n, & \text{otherwise,} \end{cases}$$

and

$$r_{n+1} = \begin{cases} r_n, & \text{if } I_{\mathcal{H}}((l_n + r_n)/2, \mu) \leq \gamma, \\ (l_n + r_n)/2, & \text{otherwise.} \end{cases}$$

Note that

$$r_n - l_n = \frac{g - f}{2^{n-1}}$$

and thus

$$\lim_{n\to\infty} l_n = \lim_{n\to\infty} r_n.$$

Let us denote this limit function by h_γ. Sequences $\{l_n\}_{n\in\mathbb{N}}$ and $\{r_n\}_{n\in\mathbb{N}}$ were constructed in a such way that $I_{\mathcal{H}}(l_n, \mu) \leq \gamma$ and $I_{\mathcal{H}}(r_n, \mu) \geq \gamma$ for all $n \in \mathbb{N}$. This implies that $I_{\mathcal{H}}(h_\gamma, \mu) \leq \gamma$ and $I_{\mathcal{H}}(h_\gamma, \mu) \geq \gamma$ and thus we found a function h_γ such that $I_{\mathcal{H}}(h_\gamma, \mu) = \gamma$ and $f \leq h_\gamma \leq g$. □

Combining previous propositions and the fact that all $f \in v$ are bounded by a_v from below and by b_v from above we obtain the main theorem of this section.

Theorem 1 *Let $\mathcal{H}$ be a decomposition system, let $v \in \mathcal{V}$ be an interval-valued function, and let $\mu \in \mathcal{M}$ be a capacity. Then*

$$\hat{I}_{\mathcal{H}}(v, \mu) = [I_{\mathcal{H}}(a_v, \mu), I_{\mathcal{H}}(b_v, \mu)]\,.$$

4 Direct Modification of Definition

A direct modification of the original definition of decomposition integrals offers another look on interval-valued decomposition integral.

Definition 2 Let $\mathcal{H}$ be a decomposition system. An interval-valued decomposition integral is an operator $\bar{I}_{\mathcal{H}} \colon \mathcal{V} \times \mathcal{M} \to [0, \infty[$ given by

$$\bar{I}_{\mathcal{H}}(v, \mu) = \bigvee_{\mathcal{D} \in \mathcal{H}} \bar{C}_{\mathcal{D}}(v, \mu)$$

where $\bar{C}_{\mathcal{D}} \colon \mathcal{V} \times \mathcal{M} \to [0, \infty[$ is an interval-valued collection integral given by

$$\bar{C}_{\mathcal{D}}(v, \mu) = \bigvee \left\{ \sum_{A \in \mathcal{D}} [\alpha_A, \beta_A] \mu(A) \colon \sum_{A \in \mathcal{D}} [\alpha_A, \beta_A] 1_A \leq v, [\alpha_A, \beta_A] \geq [0, 0] \right\}.$$

Let us start be proving a similar result as in previous section for interval-valued collection integrals.

Proposition 3 *Let $\mathcal{D}$ be a collection, let $v \in \mathcal{V}$ be an interval-valued function and let $\mu \in \mathcal{M}$ be a capacity. Then*

$$\bar{C}_{\mathcal{D}}(v, \mu) = [C_{\mathcal{D}}(a_v, \mu), C_{\mathcal{D}}(b_v, \mu)].$$

Proof To simplify writing let us denote $C_{\mathcal{D}}(a_v, \mu)$ by ρ and $C_{\mathcal{I}}(b_v, \mu)$ by θ. In the definition of $\bar{C}_{\mathcal{D}}$, let us loosen the condition $[\alpha_A, \beta_A] \geq [0, 0]$ by two conditions $\alpha_A \geq 0$ and $\beta_A \geq 0$. This would imply that $\bar{C}_{\mathcal{D}}(v, \mu) \leq [\rho, \theta]$. Now let us assume two more strict conditions. The first one is that $\alpha_A = 0$. This one implies that $\bar{C}_{\mathcal{D}}(v, \mu) \geq [0, \theta]$. The other one is that $\alpha_A = \beta_A$ which implies the inequality $\bar{C}_{\mathcal{D}}(v, \mu) \geq [\rho, \rho]$. In other words,

$$\bar{C}_{\mathcal{D}}(v, \mu) \geq [0, \theta] \vee [\rho, \rho] = [\rho, \theta]$$

and thus the proposition follows. □

Now we can prove the main theorem of this section.

Theorem 2 *Let $\mathcal{H}$ be a decomposition system, let $v \in \mathcal{V}$ be an interval-valued function, and let $\mu \in \mathcal{M}$ be a capacity. Then*

$$\bar{I}_{\mathcal{H}}(v, \mu) = [I_{\mathcal{H}}(a_v, \mu), I_{\mathcal{H}}(b_v, \mu)].$$

Proof Using the previous proposition and the properties of interval operations we obtain that

$$\bar{I}_{\mathcal{H}}(v, \mu) = \bigvee_{\mathcal{D} \in \mathcal{H}} \bar{C}_{\mathcal{D}}(v, \mu) = \bigvee_{\mathcal{D} \in \mathcal{H}} [C_{\mathcal{D}}(a_v, \mu), C_{\mathcal{D}}(b_v, \mu)]$$
$$= \left[\bigvee_{D \in \mathcal{H}} C_{\mathcal{D}}(a_v, \mu), \bigvee_{D \in \mathcal{H}} C_{\mathcal{D}}(b_v, \mu) \right] = [I_{\mathcal{H}}(a_v, \mu), I_{\mathcal{H}}(b_v, \mu)]$$

as the theorem says. □

Example 1 Imagine a company of three people—two workers and one manager. The considered space is $X = \{1, 2, 3\}$, where number 1 represents the first worker, 2 represents the second worker, and 3 represents the manager. This company consists of a working space where at most two people can work at once. Also, if manager is in the work, then he is in the work all day. This leads to the following decomposition system

$$\mathcal{H} = \Big\{ \underbrace{\{\{1,3\},\{2,3\},\{3\}\}}_{\mathcal{D}_1}, \underbrace{\{\{1\},\{2\},\{1,2\}\}}_{\mathcal{D}_2} \Big\}.$$

The capacity μ encodes the productivity per hour of different groups of people. If workers work individually, then $\mu(\{1\}) = 5$ and $\mu(\{2\}) = 6$. If they work together, then $\mu(\{1, 2\}) = 12$. On the other hand, the productivity of the manager is zero, i.e., $\mu(\{3\}) = 0$, and a combination of a manager and worker yield to increase in the production, e.g., $\mu(\{1, 3\}) = \mu(\{2, 3\}) = 7$. Working hours of the first worker are somewhere between 4 and 6 h, the second worker's are between 5 and 7 h, and the manager's are exactly 5 h, i.e., $v(1) = [4, 6]$, $v(2) = [5, 7]$ and $v(3) = [5, 5]$.

Note that, for this example, $a_v(1) = 4$, $a_v(2) = 5$, and $a_v(3) = 5$. Similarly, $b_v(1) = 6$, $b_v(2) = 7$, and $b_v(3) = 5$. Then

$$I_{\mathcal{H}}(a_v, \mu) = 50 \quad \text{and} \quad I_{\mathcal{H}}(b_v, \mu) = 72,$$

and thus $\hat{I}_{\mathcal{H}}(v, \mu) = [50, 72]$.

Now, for the second part, see that

$$\bar{C}_{\mathcal{D}_1}(v, \mu) = [35, 35] \quad \text{and} \quad \bar{C}_{\mathcal{D}_2}(v, \mu) = [50, 72]$$

and thus

$$\bar{I}_{\mathcal{H}}(v, \mu) = [35, 35] \vee [50, 72] = [50, 72].$$

In this example we simulated two different mechanisms of introduced extensions of a decomposition integral for interval-valued functions.

5 Concluding Remarks

In this contribution we introduced two different extensions of decomposition integrals for interval-valued functions. The first one is based on well-known Aumann integral that is an extension of Riemann integral for arbitrary set functions and the second one is an extension based on interval operations. Interestingly, these two at a first sight two completely different approaches are equivalent and thus lead to the same integral.

As a future research referring this work we may examine the properties of these extensions and provide some applications. Also, instead of finite space X we may consider arbitrary spaces.

References

1. Aumann, R.J.: Integrals of set-valued functions. J. Math. Anal. Appl. **12**, 1–12 (1965)
2. Bustince, H., et al.: A new approach to interval-valued Choquet integrals and the problem of ordering in interval-valued fuzzy set applications. IEEE Trans. Fuzzy Syst. **21**(6), 1150–1152 (2013)
3. Choquet, G.: Theory of capacities. Annales de l'Institut Fourier **5**, 131–295 (1954)
4. Even, Y., Lehrer, E.: Decomposition-integral: unifying Choquet and the concave integrals. Econ. Theory **56**(1), 33–58 (2014)
5. Jang, L.C.: Interval-valued Choquet integrals and their applications. J. Appl. Math. Comput. **16**, 429–443 (2004)
6. Lehrer, E.: A new integral for capacities. Econ. Theory **39**(1), 157–175 (2009)
7. Li, J., Li, S.: Aumann type set-valued Lebesgue integral and representation theorem. Int. J. Comput. Intell. Syst. **2**(1), 83–90 (2009)
8. Mesiar, R., Stupňanová, A.: Decomposition integrals. Int. J. Approx. Reason. **54**(8), 1252–1259 (2013)
9. Precupanu, A.M.: The Aumann-Gould integral. Mediterr. J. Math. **5**(4), 429–441 (2008)
10. Šeliga, A.: Decomposition integral without alternatives, its equivalence to Lebesgue integral and computational algorithms. Submitted
11. Shilkret, N.: Maxitive measure and integration. Indagationes Mathematicae **33**, 109–116 (1971)
12. Wu, C., Gong, Z.: On Henstock integrals of interval-valued functions and fuzzy-valued functions. Fuzzy Sets Syst. **115**(3), 377–391 (2000)
13. Yang, Q.: The PAN-integral on the fuzzy measure space. Fuzzy Math. **3**, 107–114 (1985)
14. Zhang, D., Wang, Z.: Fuzzy integrals of fuzzy-valued functions. Fuzzy Sets Syst. **54**(1), 63–67 (1993)

Relational Powerset Theories

Jiří Močkoř

Abstract For general powerset theories in categories, relational powerset theories in these categories are introduced. These structures are defined as categories whose objects are the original powerset objects with relational structures defined on these objects. This construction generalizes classical relations or fuzzy relations in sets and it enables to defines these relational structures on more general powerset objects based on lattice-valued fuzzy structures.

Keywords Fuzzy sets · Relational structures · Powerset theory · Category

1 Introduction

The powerset structures are widely used in algebra, logic, topology and also in computer science. The standard example of a powerset structure $P(X) = \{A : A \subseteq X\}$ and the corresponding extension of a mapping $f : X \to Y$ to the map $f_P^{\to} : P(X) \to P(Y)$ is widely used in almost all branches of mathematics and their applications, including computer science. Because the classical set theory can be considered to be a special part of the fuzzy set theory, it is natural that powerset objects associated with fuzzy sets were soon investigated as generalizations of classical powerset objects. The first approach was done by Zadeh [12], who defined $[0, 1]^X$ (or L^X, where L is an appropriate ordered structure) to be a new powerset object $Z(X)$ instead of $P(X)$ and introduced the new powerset operator $f_Z^{\to} : Z(X) \to Z(Y)$. The first real attempt to uniquely derive the powerset operator $f_Z^{\to}$ from $f_P^{\to}$ and not only explicitly stipulate them was the works of Rodabaugh [9], who gave very serious basis for further research of powerset objects and operators.

This research was partially supported by the project 18-06915S provided by the Grant Agency of the Czech Republic.

J. Močkoř (✉)
Institute for Research and Applications of Fuzzy Modeling, University of Ostrava, 30. dubna 22, 701 03 Ostrava 1, Czech Republic
e-mail: Jiri.Mockor@osu.cz

M. E. Cornejo et al. (eds.), *Computational Intelligence and Mathematics for Tackling Complex Problems 2*, Studies in Computational Intelligence 955,
https://doi.org/10.1007/978-3-030-88817-6_22

An important step in this direction was the introduction of general *powerset theory* defined by Rodabaugh [10] as a special structure describing powerset objects. The idea of using powerset theory **T** for fuzzyfication is based on extension of objects X of a category **K** to another object $T(X)$ in the category of complete semilattices, which may be regarded as an object of a "cloud of fuzzy states". Typical examples of powerset objects in structures which are derived from classical fuzzy sets are, e.g., sets $F(X,\delta) \subseteq L^X$ of L-valued fuzzy sets in a set X with a L-valued similarity relation δ, which are extensional with respect to δ (see [5]), the sets $Q(X,\mathcal{A}) \subseteq L^X$ of all F-transforms of L-valued fuzzy sets in a set X with respect to a given fuzzy partition $\mathcal{A}$ on a set X [6, 7], or the set $D(X,R) \subseteq L^X$ of all upper approximations of L-valued fuzzy sets in a set X with respect to fuzzy relation R. It can be proven that all these structures can be defined as elements of $T(X)$ for a suitable category **K** and a powerset theory **T** in **K**.

Powerset theories were used to develop fuzzy topological theories for lattice-valued mathematics. The basic idea of this approach was the definition of a new theory that contains, in addition to powerset objects also topology defined on these powerset objects, and the inclusion of relevant axioms concerning continuity of maps to this new theory (see, e.g., Rodabaugh [10], Solovyov [11] with extended list of references). This new theory was called *topological powerset theory*. This approach made it possible to apply topological methods to other powerset structures than the traditional structures $P(X)$ and $Z(X)$, and also to better understand the algebraic character of these structures, using tools from the theory of monads.

In this paper, we will focus on extending this idea to structures with relations. In that way, from a concrete category **K** and a powerset theory **T** in **K**, we define a new category $\mathbf{K}_{rel}[\mathbf{T}]$ that contains, in addition to powerset objects also relations defined on powerset objects. This allows us to use relational powerset theories in more general environment than did conventional fuzzy possibilities. In that way, structures defined over powerset objects can extend the possibilities of applying traditional fuzzy structures to other areas as well.

2 Relational Powerset Theory

For L-valued fuzzy sets defined in the set X, the powerset object L^X is in many cases the basic structure for other algebraic, order or topological structures. A typical example of this is a fuzzy topology in a set X, which is defined as a special map $\mathcal{T} : L^X \to L$. In these cases, we can then talk, for example, about the topological theory of a powerset theory. This approach was systematically used in Rodabaugh paper [10], where the concept of a topological powerset theory was introduced.

In the paper we use an analogical approach and we define the notion of a relational powerset theory. In this way, we obtain a tool that, as in the case of fuzzy relations in sets, allows to apply the theory of relations of different types in powerset objects in various categories.

We start with the basic definition of the powerset theory in a category defined by Rodabaugh [10], and then we introduce the category of relations on powerset objects, associated with a powerset theory in a category.

The following definition of Rodabaugh introduces a general notion of the $CSLAT$-powerset theory, where $CSLAT$ is the category of complete semi-lattices with semi-lattice homomorphisms.

Definition 1 Let $\mathbf{K}$ be a category. Then $\mathbf{T} = (T, \rightarrow, V, \eta)$ is called ***CSLAT*-powerset theory in K**, if

1. $T : \mathbf{K} \rightarrow CSLAT$ is an object-mapping,
2. for each morphism $f : X \rightarrow Y$ in $\mathbf{K}$, there exists a morphism $f_{\mathbf{T}}^{\rightarrow} : T(X) \rightarrow T(Y)$ in the category $CSLAT$,
3. $V : \mathbf{K} \rightarrow \mathbf{Set}$ is a concrete (or, forgetfull) functor,
4. For each object $X \in \mathbf{K}$, η determines in $\mathbf{Set}$ a mapping $\eta_X : V(X) \rightarrow T(X)$,
5. For each $f : X \rightarrow Y$ in $\mathbf{K}$, the following diagram commutes:

$$\begin{array}{ccc} V(X) & \xrightarrow{V(f)} & V(Y) \\ {\scriptstyle \eta_X}\downarrow & & \downarrow{\scriptstyle \eta_Y} \\ WT(X) & \xrightarrow{W(f_T^{\rightarrow})} & WT(Y), \end{array}$$

where $W : CSLAT \rightarrow \mathbf{Set}$ is a forgetful functor,

For simplicity, instead of $CSLAT$-powerset theory we say only powerset theory. We use the idea of L.A. Manes [2–4], who defined the notion of a T-relation on X as a morphism $X \rightarrow T(X)$, where T is a special endofunctor of a category $\mathbf{K}$. In our case, we will modify this concept of the T-relation so that instead of the endofunctor $T : \mathbf{K} \rightarrow \mathbf{K}$ we will use the functor T from a powerset theory $\mathbf{T}$ in a category $\mathbf{K}$.

Definition 2 Let $\mathbf{K}$ be a category and let $\mathbf{T} = (T, \rightarrow, V, \eta)$ be a powerset theory in $\mathbf{K}$. Then a **T-relational powerset theory** in $\mathbf{K}$ is a category $\mathbf{K}_{rel}[\mathbf{T}]$ defined by:

1. Objects are pairs (X, R), where X is an object in $\mathbf{K}$ and $R : V(X) \rightarrow T(X)$ is a morphism in $\mathbf{Set}$, such that $\eta_X \leq R$ in point-wise ordering from $T(X)$.
2. A morphism $f : (X, R) \rightarrow (Y, S)$ is a morphism $f : X \rightarrow Y$ in $\mathbf{K}$, such that in the diagram

$$\begin{array}{ccc} V(X) & \xrightarrow{V(f)} & V(Y) \\ {\scriptstyle R}\downarrow & \leq & \downarrow{\scriptstyle S} \\ T(X) & \xrightarrow{f_T^{\rightarrow}} & T(Y) \end{array}$$

the inequality $f_T^{\rightarrow}.R \leq S.V(f)$ holds, where $\leq$ is the point-wise ordering inherited from the semilattice $T(Y)$.

The category $\mathbf{K}_{rel}[\mathbf{T}]$ represents a natural generalization of the category **Rel** of sets with relations and the category **FRel** of sets with L-valued fuzzy relations. In fact, let us consider the following examples.

Example 1 Let $\mathbf{K} = \mathbf{Set}$ and let $\mathbf{P} = (P, \rightarrow, 1_{\mathbf{Set}}, \chi)$ be the classical powerset theory, i.e., $P(X) = 2^X$, $f_P^{\rightarrow}(A) = f(A)$, $\chi_X(x) = \{x\}$. Then objects of the category $\mathbf{Set}_{rel}[\mathbf{P}]$ are pairs (X, R), where R can be identified with the relation $R \subseteq X \times X$. Therefore, $\mathbf{Set}_{rel}[\mathbf{P}]$ is isomorphic to the classical category **Rel** of sets with relations, with morphisms $f : (X, R) \rightarrow (Y, S)$ such that $(x, y) \in R \Rightarrow (f(x), f(y)) \in S$. □

Example 2 Let $\mathbf{K} = \mathbf{Set}$ and let $\mathbf{Z} = (Z, \rightarrow, 1_{\mathbf{Set}}, \rho)$ be the classical Zadeh's powerset theory of L-valued fuzzy sets, where L is a complete lattice. Objects of the category $\mathbf{Set}_{rel}[\mathbf{Z}]$ are pairs (X, R), where the map $R : X \rightarrow Z(X) = L^X$ can be identified with the reflexive fuzzy relation $R \subsetapprox X \times X$ and with morphisms $f : (X, R) \rightarrow (Y, S)$, such that $R(x, y) \leq S(f(x), f(y))$. Hence, $\mathbf{Set}_{rel}[\mathbf{Z}]$ is isomorphic to the category **FRel**. □

Example 3 Let $L = (L, \vee, \wedge, \otimes, \rightarrow)$ be a complete residuated lattice and let $\mathbf{Set}(L)$ be the category with objects (X, δ), where X is a set and δ is L-valued similarity relation on X, with maps $f : X \rightarrow Y$ as morphisms $f : (X, \delta) \rightarrow (Y, \gamma)$, such that $\delta(x, y) \leq \gamma(f(x), f(y))$. Then the powerset theory $\mathbf{F} = (F, \rightarrow, V, \eta)$ in the category $\mathbf{Set}(L)$ is defined by

1. The object function $F : \mathbf{Set}(L) \rightarrow CSLAT$ is defined by $F(X, \delta) =$ the set of all extensional L-fuzzy sets in X with respect δ,
2. for each morphism $f : (X, \delta) \rightarrow (Y, \gamma)$ in $\mathbf{Set}(L)$, $f_F^{\rightarrow} : F(X, \delta) \rightarrow F(Y, \gamma)$ is defined by $f_F^{\rightarrow}(s)(y) = \bigvee_{x \in X} s(x) \otimes \gamma(f(x), y)$,
3. $V : \mathbf{Set}(L) \rightarrow \mathbf{Set}$ is the forgetfull functor,
4. for each $(X, \delta) \in \mathbf{Set}(L)$, $\eta_{(X,\delta)} : X \rightarrow F(X, \delta)$ is defined by $\eta_{(X,\delta)}(a)(x) = \delta(a, x)$, for $a, x \in X$.

The category $\mathbf{Set}(L)_{rel}[\mathbf{F}]$ is isomorphic to the category with objects $((X, \delta), R)$, where R can be identified with a reflexive L-valued relation in a set $X \times X$, such that $R(x, y) \otimes \delta(y, y') \leq R(x, y')$ holds for arbitrary $x, y, y' \in X$ and with morphisms $f : ((X, \delta), R) \rightarrow ((Y, \gamma), S)$, such that

$$\delta(x, y) \leq \gamma(f(x), f(y)),$$
$$\bigvee_{z \in X} R(x, z) \otimes \gamma(f(z), y) \leq S(f(x), y)$$

hold. It is clear that, in this case, **FRel** is a full subcategory in $\mathbf{Set}(L)_{rel}[\mathbf{F}]$, if L is a complete residuated lattice. □

In the next Proposition we introduce the powerset structure $\mathbf{D}^{\uparrow}$ in the category **FRel** of sets with reflexive L-valued fuzzy relations, where L is a complete residuated lattice. This powerset theory represents upper approximations of L-valued fuzzy with respect to a L-valued fuzzy relation.

Proposition 1 *Let* $L = (L, \vee, \wedge, \otimes, \rightarrow)$ *be a complete residuated lattice. Let* $\mathbf{D}^{\uparrow} = (D^{\uparrow}, \rightarrow, U, \rho^{\uparrow})$ *be defined in the category* **FRel** *by*

1. $D^{\uparrow}$: **FRel** $\to$ *CSLAT is the object function defined by* $D^{\uparrow}(X, R) = \{\overline{R}(s) : s \in Z(X)\} \subseteq Z(X)$, *where* $\overline{R}(s)(x) = \bigvee_{z\in X} s(z) \otimes R(z, x)$, $x \in X$.
2. $U(X, R) = X$, $U(f) = f$,
3. *For* $(X, R) \in$ **FRel**, $\rho^{\uparrow}_{(X,R)} : X \to D^{\uparrow}(X, R)$ *is defined by*

$$x \in X, \quad \rho^{\uparrow}_{(X,R)}(x) = \overline{R}(\rho_X(x)) \in D^{\uparrow}(X, R),$$

where $\rho_X(x)$ *is the characteristic function of* $\{x\}$ *in* X,
4. *For any morphism* $f : (X, R) \to (Y, S)$ *in* **FRel**, $f^{\to}_{D^{\uparrow}} : D^{\uparrow}(X, R) \to D^{\uparrow}(Y, S)$ *is defined by*

$$s \in Z(X), \quad f^{\to}_{D^{\uparrow}}(\overline{R}(s)) = \overline{S}(f^{\to}_{Z}(s))$$

Then $\mathbf{D}^{\uparrow}$ *is a* $CSLAT$*-powerset theory in the category* **FRel**.

Because the category $\mathbf{K}_{rel}[\mathbf{T}]$ is a natural generalization of the category **FRel**, the question arises whether in this general category a powerset theory can be defined that would be an extension of the powerset theory $\mathbf{D}^{\uparrow}$ in the category **FRel**. To clarify this, we show that in the category $\mathbf{K}_{rel}[\mathbf{T}]$ there is a powerset theory that represents a generalization of the powerset theory $\mathbf{D}^{\uparrow}$ in the category **FRel**.

Let us consider the following definition.

Definition 3 Let $\mathbf{T} = (T, \to, V, \eta)$ be a $CSLAT$-powerset theory in a category **K**, such that for arbitrary object $(X, R) \in \mathbf{K}_{rel}[\mathbf{T}]$ and $s \in T(X)$ there exists $s \star R \in T(X)$, and for arbitrary $\{s_i : i \in I\} \subseteq T(X)$, $(\bigvee_{i\in I} s_i) \star R = \bigvee_{i\in I}(s_i \star R)$ holds. Let the structure $\mathbf{E} = (E, \to, U, \xi)$ be defined by

1. $U : \mathbf{K}_{rel}[\mathbf{T}] \to$ **Set** is a forgetfull functor, such that $U(X, R) = V(X)$, $U(f) = V(f)$.
2. The object function $E : \mathbf{K}_{rel}[\mathbf{T}] \to CSLAT$ is defined by $E(X, R) = (\{\overline{R}(s) : s \in T(X)\}, \leq)$, where $\leq$ is the ordering in $T(X)$ and for $s \in T(X)$, $\overline{R} : T(X) \to T(X)$ is defined by $\overline{R}(s) = s \star R$, .
3. For each morphism $f : (X, R) \to (Y, S)$ in $\mathbf{K}_{rel}[\mathbf{T}]$, $f^{\to}_{E} : E(X, R) \to E(Y, S)$ is defined by

$$s \in T(X), \quad f^{\to}_{E}(\overline{R}(s)) = \overline{S}(f^{\to}_{T}(s)).$$

4. For each $x \in U(X, R)$, the map $\xi_{(X,R)} : U(X, R) \to E(X, R)$ in the category **Set** is defined by $\xi_{(X,R)}(x) = \overline{R}(\eta_X(x)) \in E(X, R)$.

Proposition 2 *The structure* **E** *is a* $CSLAT$*-powerset theory in the category* $\mathbf{K}_{rel}[\mathbf{T}]$.

The proof of this proposition will appear elsewhere.

Definition 4 The $CSLAT$-powerset theory **E** in the category $\mathbf{K}_{rel}[\mathbf{T}]$ is called the powerset theory of upper **T**-approximations.

In the following examples we show that **E** generalizes powerset theory $\mathbf{D}^{\uparrow}$ in the category **FRel**.

Example 4 Let $\mathbf{K} = \mathbf{Set}$, $\mathbf{T} = \mathbf{P}$ be the classical powerset theory in **Set** from the Example 1. Then, **P** satisfies the conditions from Definition 3 and there exists a powerset theory **E** of upper **P**-approximations in the category $\mathbf{Set}_{rel}[\mathbf{P}]$. In fact, if $(X, R) \in \mathbf{Set}_{rel}[\mathbf{P}]$, then for $A \in P(X)$ we can define

$$A \star R = R[A] = \{x \in X : \exists a \in A, (a, x) \in R\}.$$

It is clear that $R[\bigcup_i A_i] = \bigcup_i R[A_i]$ holds and $E(X, R)$ is the set of all upper **P**-approximations of subsets in X by the relation R. □

Example 5 Let L be a complete residuated lattice and let $\mathbf{K} = \mathbf{Set}$, $\mathbf{T} = \mathbf{Z}$ be from Example 2. Then, the powerset theory **Z** satisfies the conditions from the Definition 3 and there exists the powerset theory **E** of upper **Z**-approximations in the category $\mathbf{Set}_{rel}[\mathbf{Z}]$. In fact, if $(X, R) \in \mathbf{Set}_{rel}[\mathbf{Z}]$, then for $s \in Z(X)$, the fuzzy set $s \star R$ is defined by $(s \star R)(x) := \overline{R}(s)(x)$ from Proposition 1. In this case, $E(X, R)$ is a set of classical upper L-fuzzy approximations and **E** is identical to $\mathbf{D}^{\uparrow}$. □

The construction of the category $\mathbf{K}_{rel}[\mathbf{T}]$, which generalizes the category **FRel**, preserves many of the original properties of the category **K**. Among the properties preserved by this construction is the property of being a topological category (see, e.g., [1] for the definition).

Theorem 1 *Let* **K** *be a category and* $\mathbf{T} = (T, \rightarrow, V, \eta)$ *a* $CSLAT$*-powerset theory in* **K**. *If* **K** *is a topological category, the relational powerset theory* $\mathbf{K}_{rel}[\mathbf{T}]$ *is also a topological category.*

The proof will appear elsewhere.

3 Conclusions

In this paper, we have attempted to at least partially solve the problem of how it is possible to systematically extend the various structures originally defined on the set of L-valued fuzzy sets L^X to the powerset objects defined in categories, using the general powerset theory of objects introduced by Rodabaugh [9]. For this purpose, we have dealt with widely used relational structures and we have introduced the concept of the relational powerset theory in a category **K**. This theory is defined as the new category $\mathbf{K}_{rel}[\mathbf{T}]$, created from the original category **K**, the powerset theory **T** and the respective axioms of a relational theory. We proved that many special examples of relational structures defined on standard powerset objects L^X of fuzzy sets are objects of this category $\mathbf{K}_{rel}[\mathbf{T}]$. We prove that the new category preserves some properties of the original category **K**. Namely, we show that if **K** is a topological category then $\mathbf{K}_{rel}[\mathbf{T}]$ is also topological.

References

1. Herrlich, H.: Categorical topology 1971–1981. In: Proceedings of the Fifth Prague Topology Symposium, pp. 279–383. Heldermann, Berlin (1982)
2. Manes, E.G.: Algebraic Theories. Springer, Berlin (1976)
3. Manes, E.G.: A class of fuzzy theories. J. Math. Anal. Appl. **85**, 409–451 (1982)
4. Manes, L.A.: Book review "Fuzzy sets and systems, Theory and applications". Bull. (New Series) Am. Math. Soc. **7**(3) (1982)
5. Močkoř, J.: Powerset operators of extensional fuzzy sets. Iran. J. Fuzzy Syst. **15**(2), 143–163 (2018)
6. Močkoř, J., Holčapek, M.: Fuzzy objects in spaces with fuzzy partitions. Soft Comput. **21**(24), 7269–7284 (2017)
7. Perfilieva, I.: Fuzzy transforms: theory and applications. Fuzzy Sets Syst. **157**, 993–1023 (2006)
8. Rodabaugh, S.E.: Powerset operator foundation for poslat fuzzy set theories and topologies. In: Höhle, U., Rodabaugh, S.E. (eds.). Mathematics of Fuzzy Sets: Logic, Topology and Measure Theory, The Hnadbook of Fuzzy Sets Series, vol. 3, pp. 91–116. Kluwer Academic Publishers, Boston, Dordrecht (1999)
9. Rodabaugh, S.E.: Powerset operator based foundation for point-set lattice theoretic (poslat) fuzzy set theories and topologies. Quaestiones Mathematicae **20**(3), 463–530 (1997)
10. Rodabaugh, S.E.: Relationship of algebraic theories to powerset theories and fuzzy topological theories for lattice-valued mathematics. Int. J. Math. Math. Sci. **2007**, 1–71
11. Solovyov, S.A.: Powerset operator foundations for catalg fuzzy set theories. Iran. J. Fuzzy Syst. **8**(2), 1–46 (2001)
12. Zadeh, L.A.: Fuzzy sets. Inf. Control **8**, 338–353 (1965)

On Some Categories Underlying Knowledge Graphs

Ondrej Krídlo, Manuel Ojeda-Aciego, Tim Put, and Marek Z. Reformat

Abstract This paper proposes a method to provide a categorical structure for RDF-based data representing descriptions of entities. This is the first step towards our aim to further develop the underlying categorical structure so that we can eventually provide an internal logic which enables us to focus on analysis of properties of entities.

Keywords RDF · Formal concept analysis · Category theory

1 Introduction

Category theory, a branch of modern mathematics, formalizes concepts as objects and relations between them as arrows. The abstract nature of the category theory makes it an attractive tool for modeling and validating structures and connections between them in almost any area of science and engineering.

Graph-based data formats enable representation of a single entity as a graph node related to other nodes via different relations. Those relations and other nodes can be treated as properties of the entity. Analysis of such data formats can provide insight

*Partially supported by the Slovak Research and Development Agency contract No. APVV-15-0091.
**Partially supported by the Ministry of Science, Innovation, and Universities (MCIU), the State Agency of Research (AEI) and the European Social Fund (FEDER) through the research project PGC2018-095869-B-I00 (MCIU/AEI/FEDER, UE).
***Partially supported by Natural Sciences and Engineering Research Council, Canada.

O. Krídlo
Faculty of Science, Institute of Computer Science, University of P.J. Šafárik in Košice, Košice, Slovakia

M. Ojeda-Aciego (✉)
Departamento de Matemática Aplicada, Universidad de Málaga, Málaga, Spain
e-mail: aciego@uma.es

T. Put · M. Z. Reformat
Electrical and Computer Engineering, University of Alberta, Edmonton T6G 2V4, Canada

M. E. Cornejo et al. (eds.), *Computational Intelligence and Mathematics for Tackling Complex Problems 2*, Studies in Computational Intelligence 955,
https://doi.org/10.1007/978-3-030-88817-6_23

and better understanding of represented information. It would allow to discover rules that govern relationships between different entities (nodes of a graph), as well as to enable building structures managing synthesis of new information. The Resource Description Framework (RDF) data format [15] introduced by the Semantic Web as a standard for Linked Open Data is a popular graph-based data format in which each piece of data is stored as an RDF triple. Each triple contains two entities, two nodes in a graph, called a subject and an object and a relation between them, an edge in a graph, called a property. Processing data represented as knowledge graphs in an RDF format is generating a lot of attention. There are multiple works focusing on different aspects related to RDF-based knowledge graphs: from their construction [8], via storage [14], querying strategies [2, 13] and extracting information [7], to applications [11, 12], just to mention a few. The fact that subjects of triples could be also objects of another triples, and vice versa, means that we deal with a network of entities highly interconnected via properties.

In this context, it is important to learn more about relations existing between pieces of data. Once we name them, we can talk about properties and features of entities represented by the data. It means that learning more about relations and features allow us to understand more about entities and how they interact between themselves. In order to accomplish that, it is important to develop methods and techniques capable of building logical structures and frameworks reflecting relations between entities and their features. That would enable a proper interpretation of data done in an automatic way.

In this paper, a mathematical scheme for analysis of and inference about graph-based data, RDF in particular, is proposed and described. An RDF graph is a very interesting feature-based representation of pieces of data. A single node is connected to multiple other nodes via properties (relations), and those nodes together with those properties are features of this single node that can represent an entity or a concept. Therefore, we apply elements of category theory to a graph of concepts to build a structure representing relations between sets of concept features. In other words, we transform specific information into feature-based definition of concepts, and further create a set-based representation of those definitions, i.e., sets that contain features of concepts' definitions. Specifically, the categorical structure will be provided in terms of the construction of a category of isotone Galois connections arising from a set of formal contexts associated to the different properties of our given RDF data.

The contributions of the paper can be summarized as providing categorical structure to RDF data, based on the fact that a category is just a simple interpretation of a network of resources connected via properties. An example illustrating a fragment of the free construction **RdfT** is presented and, then, we construct a category of powerset complete lattices and isotone Galois connections which can be interpreted as a category of relations. Finally, we include some conclusions and prospects for future work.

Due to lack of space, we will assume that the reader knows the basic notions of formal concept analysis, namely, formal context, derivation operator, (isotone) Galois connection, and the definition of a category.

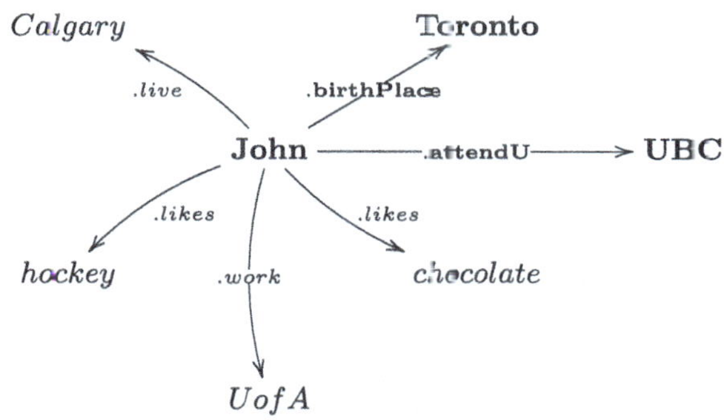

Fig. 1 Set of triples describing *John*

2 RDFData in a Nutshell

In this section, we describe fundamental aspects of RDF format necessary to introduce and explain the proposed approach of formal analysis of RDF data. Additionally, we introduce a fragment of an RDF-based knowledge graph that plays the role of a running example throughout the paper.

A single RDF-triple `<subject-property-object>` can be perceived as a feature of an entity identified by the `subject`. In other words, each single triple is treated as a feature of its `subject`. Multiple triples with the same `subject` constitute a description of a given entity. A set of few RDF triples with *John* as their subject is presented in Fig. 1. As it can be seen, each triple provides a piece of description of **John**.

Quite often a `subject` and `object` of one triple can be involved in multiple other triples, i.e., they can be `objects` or `subjects` of other triples. In such a case, descriptions of multiple entities can share features, and further some of the features can be centres of descriptions of another entities. Such descriptions can be interpreted as entities' definitions, and all interconnected triples as a network of interleaving definitions of entities.

Example 1 A simple example of multiple interconnected RDF triples is shown in Fig. 2. A number of properties are used: `<-.birthPlace->`, `<-.relative->`, `<-.attend`$_{\mathtt{HighSchool(HS)/University(U)}}$`->`,`<-.leader->`, and`<-.loc`$_{\mathtt{HS/U}}$`->` These are relations between subject/objects: `John`, `Susan`, `Paul`; `StrathconaHS`; `UofToronto`, `UBC`, and `UofCalgary`; and `Toronto`, `Okanagan`, and `Vancouver`.

3 Constructing Categories on RDF-Based Graphs

A process of constructing a category of RDF data and/or RDF-based concepts can be performed in multiple ways. Creating a category based on RDF-based knowledge graphs is relatively straightforward (Sect. 3.1). However, building a category of concepts requires determining RDF graphs representing definitions of concepts. One of such methods is identification of clusters of RDF-based descriptions of entities [3], and alternative one is to base the construction on the theory of formal concept analysis [5].

3.1 Category of RDF Triples: RdfT

The set of RDF triples represented in Fig. 2 can be easily converted into a category via the so-called free construction. The objects and subjects, i.e., `John`, `Susan`, `Paul`, `StrathconaHS`, `UofToronto`, `UofCalgary`, `UBC`, `Toronto`, `Okanagan`, and `Vancouver` become categorical objects; while the properties are morphisms. Additionally, each objects is equipped with an identity morphism $id_{\mathbf{node}}$. Let us call this category **RdfT**.

Following one of the basic properties of category theory, we can create sequences of morphisms (RDF properties) using the composition operator on multiple morphisms. For example, the two combined triples

`<Susan-relative-John-birth_place-Toronto>`

indicate that `John` – `Susan`'s relative – was born in `Toronto`, while

`<Susan-attend-UofT-located-Toronto-leader-John>`

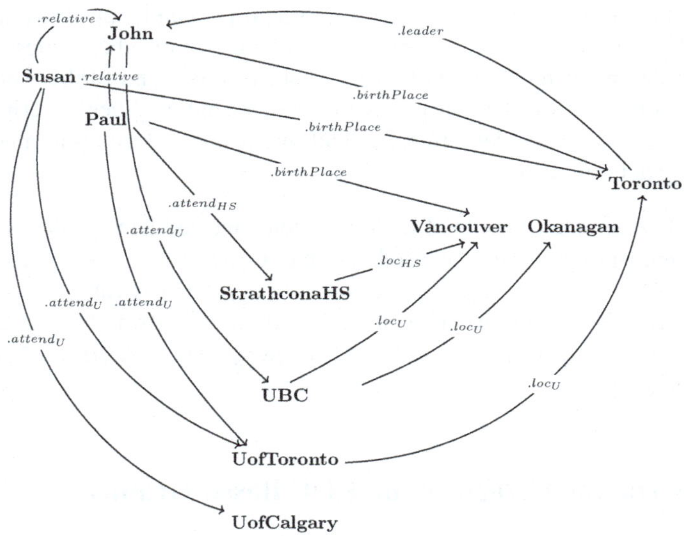

Fig. 2 Multiple RDF triples representing relations between individuals, educational institutions, and towns

says that `Susan` attends `University of Toronto`, located in `Toronto` govern by `John`. In such a way, a variety of non-trivial relations involving multiple morphisms/RDF properties can be built.

3.2 Introducing Formal Contexts and Isotone Galois Connections

Let S, P, O be sets of all assumed subjects, properties and objects, respectively, in an RDF knowledge graph, say T. Given any property $p \in P$, we can consider its domain and its range, namely,

$$S_p = \{s \in S \mid \exists o \in O \textit{ such that } \langle s, p, o\rangle \in T\}$$
$$O_p = \{o \in O \mid \exists s \in S \textit{ such that } \langle s, p, o\rangle \in T\}$$

We can also define the relation $R_p = \{(s, o) \in S_p \times O_p \mid \langle s, p, o\rangle \in T\}$; then, for any property $p \in P$ we have a formal context (S_p, O_p, R_p).

Now, we can build a category whose objects are of the form S_p and O_p for any property $p \in P$. The basic morphisms in the category will be the relations $R_p \subseteq S_p \times O_p$ but, together with the standard equality relations (which play the role of identity morphisms) and the standard composition of relations.

3.3 Construction of RDF Power Set Category

Given a property $p \in P$ and its corresponding formal context $\langle S_p, O_p, R_p\rangle$, we can define derivation operators $\uparrow_p : \mathcal{P}(S_p) \to \mathcal{P}(O_p)$ and $\downarrow_p : \mathcal{P}(O_p) \to \mathcal{P}(S_p)$ given for all $X \subseteq S_p$ and $Y \subseteq O_p$ as follows:

$$\uparrow_p (X) = \{o \in O_p \mid \exists s \in X \text{ such that } (s, o) \in R_p\}$$
$$\downarrow_p (Y) = \{s \in S_p \mid \text{ if } (s, o) \in R_p \text{ then } o \in Y\}$$

The pair $(\uparrow_p, \downarrow_p)$ forms an isotone Galois connection between the complete lattices given by the powersets $(\mathcal{P}(S_p), \subseteq)$ and $(\mathcal{P}(O_p), \subseteq)$, namely, the following equivalence holds for all $X \subseteq S_p$ and $Y \subseteq O_p$:

$$\uparrow_p(X) \subseteq Y \iff X \subseteq \downarrow_p(Y) .$$

As the composition of isotone Galois connections forms a new isotone Galois connection, we have a new category of powersets and "nice" mappings between them.

The interpretation of mappings $\uparrow$ in graph-based terms is the following: given a set of vertices X then $\uparrow(X)$ is the set of vertices adjacent to any of the vertices in X.

Obviously, as composition of mappings $\uparrow_i : \mathcal{P}(S_p^i) \to \mathcal{P}(O_p^i)$ for $i \in \{1, 2\}$, where $S_p^2 = O_p^1$ we consider the standard mapping composition, then $\uparrow_2 (\uparrow_1 (X_1))$ is the set of all vertices at distance 2 from any of the vertices in the set X_1.

Formally, we obtain the following category:

- Objects are powerset complete lattices $(\mathcal{P}(X), \subseteq)$ where X is equal to S_p or O_p for some $p \in P$
- Arrows are isotone Galois connections $(\uparrow_R, \downarrow_R)$ between powerset complete lattices $(\mathcal{P}(X), \subseteq)$ and $(\mathcal{P}(Y), \subseteq)$ for any relation $R \subseteq X \times Y$.

The category of isotone Galois connections between powerset complete lattices is equivalent to category **Rels** of relations. This can be seen as a special case of the construction in [6] in that every relation $R \subseteq X \times Y$ can be seen as a bond between formal contexts $\langle X, X, (X \times X) \setminus \{(x, x) \mid x \in X\}\rangle$ and $\langle Y, Y, (Y \times Y) \setminus \{(y, y) \mid y \in Y\}\rangle$ such that their concept lattice contains all subsets hence is isomorphic to the powerset complete lattice. In [6], a bijection can be found between all relations (bonds) between sets (special contexts) and all isotone Galois connections between powerset (concept) lattices.

4 Conclusions and Future Work

We have initiated a new attempt to provide categorical insight to the theory of RDF data; early attempts can be found in [4, 9, 10].

The construction of the category of property-based formal contexts together with isotone Galois connections [6] paves the way to a further interaction between results in Formal Concept Analysis and RDF. An interesting by-product of this interaction is that the needed results are well known in the fuzzy generalization of Formal Concept Analysis and, possibly, greater level of generality can be obtained, for instance, by using the Chu construction [1].

This is just another step in our long-term goal to provide additional categorical constructs in this setting in order to be able to further develop its internal logic.

References

1. Antoni, L., Cabrera, I.P., Krajči, S., Krídlo, O., Ojeda-Aciego, M.: The Chu construction and generalized formal concept analysis. Int. J. Gen. Syst. **46**(5), 458–474 (2017)
2. Arnaout, H., Elbassuoni, S.: Effective searching of RDF knowledge graphs. J. Web Semant. **48**, 66–84 (2018)
3. Chen, J.X., Reformat, M.Z.: Modelling of experienced-based data in linked data environment. In: Proceedings of Intelligent Networking and Collaborative Systems (INCoS), pp. 731–736 (2014)
4. Chen, J.X., Reformat, M.Z.: Learning categories from linked open data. In: Proceedings of Information Processing and Management of Uncertainty in Knowledge-Based Systems.

(IPMU). Communications in Computer and Information Science, vol. 444, pp. 396–405. Springer, Cham (2014)
5. Ganter, B., Wille, R.: Formal Concept Analysis—Mathematical Foundations. Springer (1999)
6. Krídlo, O., Ojeda-Aciego, M.: Revising the link between L-Chu correspondences and completely lattice L-ordered sets. Ann. Math. Artif. Intell. **72**, 91–113 (2014)
7. Lösch, U., Bloehdorn, S., Rettinger, A.: Graph Kernels for RDF Data. In: Proceedings of ESWC, Lecture Notes in Computer Science, vol. 7295, pp. 134–148 (2012)
8. Paulheim, H.: Knowledge graph refinement: a survey of approaches and evaluation methods. Semant. Web **8**(3), 489–508 (2017)
9. Reformat, M.Z., D'Aniello, G., Gaeta, M.: Knowledge graphs, category theory and signatures. In: Proceeding of the IEEE/WIC/ACM International Conference on Web Intelligence, pp. 480–487 (2018)
10. Reformat, M.Z., Put, T.: Constructing topos from RDF data. In: Proceeding of the IEEE/WIC/ACM International Conference on Web Intelligence and Intelligent Agent Technology, pp. 451–456 (2015)
11. Ruan, T., Xue, L., Wang, H., Hu, F., Zhao, L., Ding, J.: Building and exploring an enterprise knowledge graph for investment analysis. In: Proceeding of ISWC, Lecture Notes in Computer Science, vol. 9982, pp. 418–436 (2016)
12. Szekely, P., et al.: Building and using a knowledge graph to combat human trafficking. In: Proceeding of ISWC 2015, Lecture Notes in Computer Science, vol. 9367, pp. 205–221 (2015)
13. Verborgh, R., et al.: Triple pattern fragments: a low-cost knowledge graph interface for the web. J. Web Semant. **37–38**, 184–206 (2016)
14. Wang, Q., Mao, Z., Wang, B., Guo, L.: Knowledge graph embedding: a survey of approaches and applications. IEEE Trans. Knowl. Data Eng. **29**(12), 2724–2743 (2017)
15. RDF: http://www.w3.org/RDF/ (accessed June 9th, 2019)

On Two Categories of Many-Level Fuzzy Morphological Spaces

Alexander Šostak and Ingrida Uljane

Abstract We introduce many-level versions of the basic concepts of mathematical morphology, i.e. erosion and dilation, thus allowing to consider them not as a result, but as a process. We introduce two alternative many-level kinds of fuzzy morphological spaces. Fuzzy morphological spaces and reasonably defined continuous transformations of such spaces lead us to two categories, the study of which is initiated in the paper.

Keywords Many-level fuzzy relation · Many-level fuzzy relational erosion and dilation · Many-level fuzzy morphological spaces

1 Introduction

Mathematical morphology has its origins in geological problems centered in the processes of erosion and dilation. The founders of mathematical morphology are a civil engineer J. Serra [20] and a mining engineer G. Matheron [18]. At present mathematical morphology can be characterized as a mathematical theory that focuses on studying transformations, especially non-linear, of geometrical objects. Mathematical morphology has applications in different areas of theoretical and applied science, in particular in pattern recognition, image processing, digital topology, etc. Problems

The authors are thankful for the partial financial support from the project No Lzp-2018/2-0338. The first named author expresses his gratitude to Manuel Ojeda-Aciego for drawing his attention to the topic of Mathematical Morphology. Both authors are thankful to the two anonymous referees for reading the paper carefully and making constructive critical remarks.

A. Šostak (✉) · I. Uljane
Institute of Mathematics, CS University of Latvia, Riga LV-1459, Latvia
e-mail: aleksandrs.sostaks@lumii.lv

I. Uljane
e-mail: ingrida.uljane@lu.lv

Department of Mathematics, University of Latvia, Riga LV-1004, Latvia

M. E. Cornejo et al. (eds.), *Computational Intelligence and Mathematics for Tackling Complex Problems 2*, Studies in Computational Intelligence 955,
https://doi.org/10.1007/978-3-030-88817-6_24

related to mathematical morphology were studied by many authors, see e.g. [3, 4, 6, 7, 19], just to mention some of them. As far as we know, all researchers treated basic concepts of mathematical morphology, erosion and dilation, in a static way, that is as if the process has finished. On the other hand, we think that it is important also to view erosion and dilation as processes depending on some parameter interpreted, for example, as time or level of the process. It is the purpose of this paper to initiate the study of basic structures of mathematical morphology in the process of their development.

There are several different approaches to the subject of mathematical morphology. We stick here on the so called L-fuzzy relational mathematical morphology, see, e.g. [19], however to achieve our merits, instead of ordinary L-fuzzy relations, we base on *many-level L-fuzzy relations*. Besides, as different from the papers on fuzzy morphology known to us, we develop our approach in the framework of the category theory.

The paper consists of five sections. In the first one we introduce basic concepts concerning many-level L-fuzzy relations. The second section deals with upper and lower images and preimages of many-level L-fuzzy relations; these images and preimages are closely related to the operators of many-level erosion and dilation introduced and studied in Sect. 4. The fourth, main sections consists of three subsections. In the first two many-level fuzzy erosion and dilation operators are applied in order to define two kinds of many-level fuzzy morphological spaces: erosion spaces and dilation spaces. We define continuous transformation of such spaces coming thus to two categories of many-level fuzzy morphological spaces. In the third subsection we briefly discuss the approach allowing to combine erosion and dilation operators in a united structure. In the last, Conclusion section, we list some directions where the research of many-level fuzzy morphological spaces could be continued.

2 Many-Level L-fuzzy Relations

Let $L = (L, \leq_L, \wedge_L, \vee_L, *)$ be a commutative complete lattice monoid with bottom and top elements 0_L and 1_L respectively, see e.g. [13]. A complete commutative lattice monoid is called integral if $a * 1_L = a$ for all $a \in L$. Binary operation $*$ is called lower semicontinuous if $(\bigvee_i \alpha_i) * \beta = \bigvee_i (\alpha_i * \beta)$ for all $\{\alpha_i \mid i \in I\} \subseteq L$ and all $\beta \in L$. Further, let $\mapsto: L \times L \to L$ be the residuum related to operation $*$ via the Galois connection, that is $a * b \leq c \Longleftrightarrow a \leq b \mapsto c$ for all $a, b, c \in L$. In this work, $L = (L, \leq_L, \wedge_L, \vee_L, *)$ is always an integral commutative complete lattice monoid, operation $*$ is lower semicontinuous and $\mapsto$ is the corresponding residuum. Further, let $M = (M, \leq_M, \wedge_M, \vee_M)$ be a complete infinitely distributive lattice.

Definition 1 *(cf e.g.* [25]*)* A many-level (or M-level if we want to specify lattice M) L-fuzzy relation from a set X to a set Y is a mapping $R : X \times Y \times M \to L$.

R is called non-decreasing if $\alpha \le \beta$ implies $R(x, y, \alpha) \le R(x, y, \beta)$ for all $x \in X y \in Y$, $\alpha, \beta \in M$.

R is called upper semicontinuous if $R\left(x, y, \bigvee_{i \in I} \alpha_i\right) = \bigvee_{i \in I} R(x, y, \alpha_i)$ for all $x \in X$, $y \in Y$, $\{\alpha_i \mid i \in I\} \subseteq M$.

An M-level L-fuzzy relation R from X to Y R is called lower semicontinuous if $R\left(x, y, \bigwedge_{i \in I} \alpha_i\right) = \bigwedge_{i \in I} R(x, y, \alpha_i)$ for all $x \in X$, $y \in Y$, $\{\alpha_i \mid i \in I\} \subseteq M$.

Definition 2 *(cf e.g.* [21, 22, 25]*)* A many level L-fuzzy relation R from X to Y is called left connected if $\bigwedge_{y \in Y} \bigvee_{x \in X} R(x, y, \alpha) = 1$ for all $\alpha \in M$. If for every $y \in Y$ there exists $x \in X$ such that $R(x, y, \alpha) = 1_L$ for all $\alpha \in M$, then R is called *strongly left connected.* A many-level L-fuzzy relation R from X to Y is called right connected if $\bigwedge_{x \in X} \bigvee_{y \in Y} R(x, y, \alpha) = 1$ for all $\alpha \in M$. If for every $x \in X$ there exists $y \in Y$ such that $R(x, y, \alpha) = 1_L$ for all $\alpha \in M$, then R is called strongly right connected

3 Image and Preimage Operators on L-fuzzy Powersets Induced by Many-Level L-fuzzy Relations

The subject of this section are, what we call, upper and lower image and preimage operators induced by a many-valued L-fuzzy relation $R : X \times Y \times M \to L$. As we will see, they are closely related to the operators of many-level fuzzy erosion and dilation, which are of the main interest in this paper. These operators and their basic properties can be found in different papers where L-fuzzy powersets are involved For reader convenience we briefly discuss them here.

Definition 3 The upper image $R^{\to}(A) \in L^{Y \times M}$ of the fuzzy set $A \in L^X$ under the relation $R : X \times Y \times M \to L$ is defined by $R^{\to}(A)(y, \alpha) = \bigvee_x R(x, y, \alpha) * A(x)$ for all $A \in L^X$, $y \in Y, \alpha \in M$.

Definition 4 The upper preimage $R^{\leftarrow}(B) \in L^{X \times M}$ of the L-fuzzy set $B \in L^Y$ under relation $R : X \times Y \times M \to L$ is defined by $R^{\leftarrow}(B)(x, \alpha) = \bigvee_y R(x, y, \alpha) * B(y)$ for all $B \in L^Y$, $x \in X$, and $\alpha \in M$.

Definition 5 The lower image of an L-fuzzy set $A \in L^X$ is the L-fuzzy set $R^{\Rightarrow}(A) \in L^{Y \times M}$ defined by $R^{\Rightarrow}(A)(y, \alpha) = \bigwedge_{x \in X} R(x, y, \alpha) \mapsto A(x)$ for all $A \in L^X$, $y \in Y, \alpha \in M$.

Definition 6 The lower preimage of an L-fuzzy set $B \in L^Y$ is the L-fuzzy set $R^{\Leftarrow}(B) \in L^{X \times M}$ defined by $R^{\Leftarrow}(B)(x, \alpha) = \bigwedge_{y \in Y} R(x, y, \alpha) \mapsto B(y)$ for all $B \in L^Y$, $y \in Y, \alpha \in M$.

Proposition 1 *If relation $R : X \times Y \times M \to L$ is strongly left connected, then $R^{\Rightarrow}(A) \le R^{\to}(A)$ for every $A \in L^X$.*

Proof Given $y \in Y$, we take x_y such that $R(x_y, y, \alpha) = 1_L$. Then
$\bigwedge_{x \in X} R(x, y, \alpha) \mapsto A(x) \le R(x_y, y, \alpha) \mapsto A(x_y) = A(x_y)$. On the other hand, $\bigvee_{x \in X} R(x, y, \alpha) * A(x) \ge R(x_y, y, \alpha) * A(x_y) = A(x_y)$. □

Proposition 2 *If relation $R : X \times Y \times M \to L$ is strongly right connected, then $R^{\Leftarrow}(B) \leq R^{\leftarrow}(B)$ for every $B \in L^Y$.*

Proof Let $x \in X$ be fixed and let $y_x \in Y$ satisfy $R(x, y_x, \alpha) = 1_L$. Then $R^{\leftarrow}(B)(x, \alpha) = \bigvee_y (R(x, y, \alpha) * B(y) \geq R(x, y_x, \alpha) * B(y_x) = B(y_x)$. In its turn, $R^{\Leftarrow}(B)(x, \alpha) = \bigwedge_y (R(x, y, \alpha) \mapsto B(y)) \leq R(x, y_x, \alpha) \to B(y_x) = B(y_x)$. □

4 Operators of Many-Level Fuzzy Erosion and Dilation

4.1 Many-Level Fuzzy Erosion

Extending the fuzzy relational definition of erosion (see e.g. [19]) to the case of many-level L-fuzzy relations, we come to the following definition:

Definition 7 Given $A \in L^X$, its many-level erosion $\varepsilon_R(A) \in L^{Y \times M}$ is defined by $\varepsilon_R(A)(y, \alpha) = \bigwedge_{x \in X} (R(x, y, \alpha) \mapsto A(x))$. Considering erosion for all $A \in L^X$, we get the operator of erosion $\varepsilon_R : L^X \to L^{Y \times M}$.

Thus many-level fuzzy erosion ε_R is actually the lower image operator $R^{\Rightarrow}$ induced by the many-level L-fuzzy relation $R : X \times Y \times M \to L$. In the next proposition we collect some properties of erosion operator.

Proposition 3 (1) *Let $a_X : X \to L$ denote a constant function with value $a \in L$ and $\alpha \in M$. Then $\varepsilon_R(1_X, \alpha) = 1_Y$. If R is left connected, then $\varepsilon_R(a_X, \alpha) = a_Y$ for every $a \in L$.*
(2) *Operator $\varepsilon_R : L^X \to L^{Y \times M}$ is non-decreasing, that is $A_1 \leq A_2 \in L^X$ implies $\varepsilon_R(A_1) \leq \varepsilon_R(A_2)$.*
(3) *Given a family $\{A_i \mid i \in I\} \subseteq L^X$, we have $\varepsilon_R\left(\bigwedge_{i \in I} A_i\right) = \bigwedge_{i \in I} \varepsilon_R(A_i)$.*
(4) *If $\alpha \leq \beta$, $\alpha, \beta \in M$, then $\varepsilon_R(A)(y, \alpha) \geq \varepsilon_R(A)(y, \beta)$ for all $A \in L^X$, $y \in Y$.*
(5) *If relation R is upper semicontinuous, then $\varepsilon_R(A)(y, \bigvee_i \alpha_i) = \bigwedge_i \varepsilon_R(A)(y, \alpha_i)$ for all $A \in L^X$, $y \in Y$ and $\{\alpha_i \mid i \in I\} \subseteq M$.*

The proof of the first four items is easy and can be found in the literature. For completeness we prove the last property:
(5) $\varepsilon_R(A, \bigvee_i \alpha_i) = R^{\Rightarrow}(A)(y, \bigvee_i \alpha_i) = \bigwedge_x \left(R(x, y, \bigvee_i \alpha_i) \mapsto A(x)\right) = \bigwedge_x \left(\bigvee_i R(x, y, \alpha_i) \mapsto A(x)\right) = \bigwedge_x \bigwedge_i (R(x, y, \alpha_i) \mapsto A(x)) = \bigwedge_i \varepsilon_R(A, \alpha_i)$. □

4.2 Many-Level Fuzzy Dilation

Extending the fuzzy relational definition of dilation (see e.g. [19]) to the case of many-level L-fuzzy relations, we come to the following definition:

Definition 8 Given $B \in L^Y$, its many-level dilation $\delta_R(B) \in L^{X\times M}$ is defined by $\delta_R(B)(x,\alpha) = \bigvee_{y\in Y} R(x,y,\alpha) * B(y)$. Considering dilation for all $B \in L^Y$, we get the operator of dilation $\delta_R : L^Y \to L^{X\times M}$.

Thus dilation operator is actually an upper L-fuzzy preimage operator induced by the many-level L-fuzzy relation $R : X \times Y \times M \to L$.

In the next proposition we collect some properties of dilation operator.

Proposition 4 (1) *Let $b_Y : Y \to L$ denote a constant function with value $b \in L$ and $\alpha \in M$. Then $\delta_R(0_Y, \alpha) = 0_X$. If R is right connected, then $\delta_R(b_Y, \alpha) = b_X$ for any $b \in L$.*

(2) *Operator $\delta_R : L^Y \to L^X$ is non-decreasing, that is $B_1 \leq B_2 \in L^Y$ implies $\delta_R(B_1) \leq \delta_R(B_2)$.*

(3) *Given a family of L-fuzzy sets $\{B_i \mid i\in I\} \subseteq L^Y$, we have $\delta_R(\bigvee_i B_i) = \bigvee_i \delta_R(B_i)$.*

(4) *$\alpha \leq \beta$ implies $\delta_R(B)(x,\alpha) \leq \delta_R(B)(x,\beta)$*

(5) *If relation R is upper semicontinuous, then $\delta_R(B)(x, \bigvee_i \alpha_i) = \bigvee_i \delta_R(B)(x,\alpha_i)$ for any $B \in L^Y$, $\{\alpha_i \mid i \in I\} \subseteq M$.*

The proof of the first four items is easy and can be found in the literature. For completeness we prove the last property.

(5) $\delta_R(B)(x, \bigvee_i \alpha_i) = R^{\leftarrow}(x, \bigvee_i \alpha_i) = \bigvee_y (R(x,y,\bigvee_i \alpha_i) * B(y) = \bigvee_y(\bigvee_i(R(x,y,\alpha_i) * B(y) = \bigvee_i R^{\leftarrow}(B)(x,\alpha_i) = \bigvee_i \delta_R(B)(x,\alpha_i)$. □

5 Categories of Many-Level Fuzzy Morphological Spaces

In this section, we develop a categorical approach to the operators of erosion and dilation. Concerning terminology used here, the reader is referred to any standard book on category theory, for example [15, 16] or [1]. As in many other cases, the categorical viewpoint on the operators of fuzzy morphology in some situations allows to get a deeper understanding as well as permits to introduce "correct" definitions for the constructions where these operators are involved. The section consists of three subsections. In the first two we consider the category **E** of many-level fuzzy erosion spaces and the category **Δ** of many fuzzy dilation spaces. In the last two subsection we briefly discuss category of spaces obtained by combination of these operators: categories $\mathbf{E}\mathbf{\Delta}^{-1}$ and $\mathbf{E}^{-1}\mathbf{\Delta}$. Throughout the section, lattice monoid L and lattice M are fixed. In the definition of all these categories we will rely on the use the category $\mathbf{RelSet}^2$.

Definition 9 *(cf e.g.* [15, 16]*)* The objects of $\mathbf{RelSet}^2$ are triples (X, Y, R) where X, Y are sets and $R : X \times Y \times M \to L$ is an M-level L-fuzzy relation. The morphisms in $\mathbf{RelSet}^2$ are pairs of $(\varphi, \psi) : (X_1, Y_1, R_1) \to (X_2, Y_2, R_2)$ where $\varphi : X_1 \to X_2$, $\psi : Y_1 \to Y_2$ are mappings of sets and $R_1(x_1, y_1, \alpha) \leq R_2(\varphi(x_1), \psi(y_1), \alpha)$ for all $x_1 \in X_1, y_1 \in Y_1, \alpha \in M$.

5.1 Category E of Many-Valued Erosion Spaces

Definition 10 A many-level fuzzy erosion space, or just an erosion space for short is a tuple (X, Y, R, ε_R) where X, Y are sets, $R : X \times Y \times M \to L$ is an M-level L-relation and $\varepsilon_R : L^X \to L^{Y\times M}$ is the erosion operator induced by R.

In order to view erosion spaces (X, Y, R, ε_R) as a category we must specify its morphisms. We do it in the next definition justified by the topological interpretation of operator ε_R given in Sect. 5.3.

Definition 11 A continuous transformation from a space $(X_1, Y_1, R_1, \varepsilon_{R_1})$ to a space $(X_2, Y_2, R_1, \varepsilon_{R_2})$ is a pair (φ, ψ) of mappings, $\varphi : X_1 \to X_2$, $\psi : Y_1 \to Y_2$, satisfying the following conditions:

(1) $(\varphi, \psi) : (X_1, Y_1, R_1) \to (X_2, Y_2, R_2)$ is a morphism in **RelSet²**;
(2) $\varepsilon_{R_2}(\varphi(A)) \le \psi\left(\varepsilon_{R_1}(A)\right)$ $\forall A \in L^{X_1}$.

Given three erosion spaces $(X_1, Y_1, R_1, \varepsilon_{R_1})$, $(X_2, Y_2, R_2, \varepsilon_{R_2})$, $(X_3, Y_3, R_3, \varepsilon_{R_3})$ and continuous transformations $(\varphi, \psi) : (X_1, Y_1, R_1, \varepsilon_{R_1}) \to (X_2, Y_2, R_2, \varepsilon_{R_2})$, $(\varphi', \psi') : (X_2, Y_2, R_2, \varepsilon_{R_2}) \to (X_3, Y_3, R_3, \varepsilon_{R_3})$, we define their composition as $(\varphi' \circ \varphi, \psi' \circ \psi) : (X_1, Y_1, R_1, \varepsilon_{R_1}) \to (X_3, Y_3, R_3, \varepsilon_{R_3})$.

Theorem 1 *$(\varphi' \circ \varphi, \psi' \circ \psi) : (X_1, Y_1, R_1, \varepsilon_{R_1}) \to (X_3, Y_3, R_3, \varepsilon_{R_3})$ is a continuous transformation.*

Proof Note first that if $(\varphi, \psi) : (X_1, Y_1, R_1) \to (X_2, Y_2, R_2)$ and $(\varphi', \psi') : (X_2, Y_2, R_2) \to (X_3, Y_3, R_3)$ are morphisms in **RelSet²**., then their composition $(\varphi' \circ \varphi, \psi' \circ \psi) : (X_1, Y_1, R_1) \to (X_3, Y_3, R_3)$ is a morphism in **RelSet**^{2}. Further, given $A \in L^{X_1}$ we have: $\varepsilon_{R_3}(\varphi \circ \varphi')(A) = \varepsilon_{R_3}(\varphi'(\varphi(A))) \le \psi'(\varepsilon_{R_2}(\varphi(A))) \le \psi'(\psi(\varepsilon_{R_1}(A))) = (\psi' \circ \psi)(\varepsilon_{R_1}(A))$ □

Noticing that $(id_X, id_Y) : (X, Y, R, \varepsilon_R) \to (X, Y, R, \varepsilon_R)$, where $id_X : X \to X$ and $id_Y : Y \to Y$ are the identity mappings, is obviously a continuous transformation, we get the following corollary:

Corollary 1 *Erosion spaces and their continuous transformation form a category. We denote this category by* **E**, *or in case when we need to specify lattices L and M, by* **E**(LM).

In order to illustrate the benefits of the categorical approach to the study of erosion operators we prove the existence of products and coproducts in the category **E**.

Theorem 2 *Let $\{(X_i, Y_i, R_i, \varepsilon_{R_i}) \mid i \in I\}$ be a family of erosion spaces. Further, let $X = \prod_{i\in I} X_i$, $Y = \prod_{i\in I} Y_i$, be the products of the sets, let $p_i : X \to X_i$, $q_i : Y \to Y_i$ be the corresponding projections and let $R : X \times Y \times M \to L$ be defined by $R(x, y, \alpha) = \bigwedge_i R_i(x_i, y_i, \alpha)$. Then the erosion space (X, Y, R, ε_R) is the product of the family $\{X_i, Y_i, R_i, \varepsilon_{R_i} \mid i \in I\}$.*

Proof To make the proof more visible, we restrict to the case of the product of two erosion spaces $(X_1, Y_1, R_1, \varepsilon_{R_1})$, $(X_2, Y_2, R_2, \varepsilon_{R_2})$.

First, we have to prove that $(p_i, q_i) : (X, Y, R, \varepsilon_R) \to (X_i, Y_i, R_i, \varepsilon_{R_i})$ is a continuous transformation. Without loss of generality, we assume that $i = 1$ and hence we have to prove that $\varepsilon_{R_1}(p_1(A)) \le q_1(\varepsilon_R(A)$ for every $A \in L^{X_1\times X_2}$. We fix some $\bar{y}_1 \in Y_1$. Then $\varepsilon_{R_1}(p_1(A))(\bar{y}_1, \alpha) = \bigwedge_{x_1\in X_1}(R_1(x_1, \bar{y}_1, \alpha) \mapsto p_1(A)(x_1) = \bigwedge_{x_1\in X_1}(R_1(x_1, \bar{y}_1, \alpha) \mapsto \bigvee_{x_2\in X_2}(A)(x_1, x_2) = \bigwedge_{x_1\in X_1}\bigwedge_{x_2\in X_2}(R_1(x_1, \bar{y}_1, \alpha) \mapsto A$ $(x_1, x_2) = \bigwedge_{x\in X}(R_1(x_1, \bar{y}_1, \alpha) \mapsto A(x)$.

On the other hand,
$q_1(\varepsilon_R(A)(\bar{y}_1, \alpha) = \bigvee_{y_2\in Y_2}\varepsilon_R(A)(\bar{y}_1, y_2, \alpha) = \bigvee_{y_2\in Y_2}(\bigwedge_{x\in X}(R(x, y, \alpha) \mapsto A(x))$
$= \bigvee_{y_2\in Y_2}\bigwedge_{x\in X}(R_1(x_1, \bar{y}_1, \alpha) \wedge R_2(x_2, y_2) \mapsto A(x)) \ge \bigwedge_{x\in X} R_1(x_1, \bar{y}_1, \alpha) \mapsto A(x)$.
Thus $\varepsilon_{R_1}(p_1(A)) \le q_1(\varepsilon_R(A)$ for every $A \in L^{X_1\times X_2}$.

Further, let $(Z, U, R_0, \varepsilon_{R_0})$ be an erosion space and let $(\varphi_i, \psi_i) : (Z, U, R_0, \varepsilon_{R_0}) \to (X_i, Y_i, R_i, \varepsilon_{R_i})$ be continuous transformations for i=1,2. Further, let $\varphi = (\varphi_1, \varphi_2) : Z \to X_1 \times X_2$ and $\psi = (\psi_1, \psi_2) : U \to Y_1 \times Y_2$ be products of mappings. Then from the construction it is easy to see that $(\varphi, \psi) : (Z, U, R_0, \varepsilon_{R_0}) \to (X, Y, R, \varepsilon_R)$ is a continuous transformation and $(p_i, q_i) \circ (\varphi, \psi) = (\varphi_i, \psi_i) : (Z, U, R_0, \varepsilon_{R_0}) \to (X_i, Y_i, R_i, \varepsilon_{R_i})$ for $i = 1, 2$. □

By Freyd theorem (see e.g. [15, 18.22]) we have the following corollary:

Corollary 2 *Coproducts exist in category* **E**.

5.2 *Category Δ of Many-Valued Fuzzy Dilation Spaces*

Definition 12 A many-level fuzzy dilation space, or just a dilation space for short, is a tuple (X, Y, R, δ_R) where X, Y are sets, $R : X \times Y \times M \to L$ is an M-level L-fuzzy relation and $\delta_R : L^Y \to L^X$ is the dilation operator.

To view many-level fuzzy dilation spaces (X, Y, R, δ_R) as a category, we must specify its morphisms. We do it in the next definition justified by the topological interpretation of operator δ_R given in Sect. 5.3.

Definition 13 A continuous transformation from a space $(X_1, Y_1, R_1, \delta_{R_1})$ to a space $(X_2, Y_2, R_1, \delta_{R_2})$ is a pair (φ, ψ) where $\varphi : X_1 \to X_2$, $\psi : Y_1 \to Y_2$ are mappings of sets satisfying the following conditions:

(1) $(\varphi, \psi) : (X_1, Y_1, R_1) \to (X_2, Y_2, R_2)$ is a morphism in **RelSet²**;
(2) $\delta_R(\varphi(B)) \ge \psi(\delta_R(B))$ for all $B \in L^Y$;

Given three dilation spaces $(X_1, Y_1, R_1, \delta_{R_1}), (X_2, Y_2, R_2, \delta_{R_2}), (X_3, Y_3, R_3, \delta_{R_3})$ and continuous transformations $(\varphi, \psi) : (X_1, Y_1, R_1, \delta_{R_1}) \to (X_2, Y_2, R_2, \delta_{R_2})$, $(\varphi', \psi') : (X_2, Y_2, R_2, \delta_{R_2}) \to (X_3, Y_3, R_3, \delta_{R_3})$, we define their composition as $(\varphi' \circ \varphi, \psi' \circ \psi) : (X_1, Y_1, R_1, \delta_{R_1}, \delta_{R_1}) \to (X_3, Y_3, R_3, \delta_{R_3}, \delta_{R_3})$.

Theorem 3 *$(\varphi' \circ \varphi, \psi' \circ \psi) : (X_1, Y_1, R_1, \delta_{R_1}) \to (X_3, Y_3, R_3, \delta_{R_3})$ is a continuous transformation.*

Proof Given $B \in L^{Y_1}$ we have: $\delta_{R_3}(\varphi' \circ \varphi)(B) = \delta_{R_3}(\varphi'(\varphi(B))) \geq \psi'(\delta_{R_2}(\varphi(B))) \geq \psi'(\psi(\delta_{R_1}(B))) = (\psi' \circ \psi)(\delta_{R_1}(B))$. □

Noticing that $(id_X, id_Y) : (X, Y, R, \delta_R) \to (X, Y, R, \delta_R)$ where $id_X : X \to X$ and $id_Y : Y \to Y$ are the identity mappings, is obviously a continuous transformation, we get the following corollary:

Corollary 3 *Dilation spaces and their continuous transformations form a category. We denote this category by $\mathbf{\Delta}$, or in case when we need to specify lattices L and M, by $\mathbf{\Delta}(LM)$.*

Theorem 4 *Let $\{(X_i, Y_i, R_i, \delta_{R_i}) \mid i \in I\}$ be a family of dilation spaces. Further, let $X = \coprod_{i\in I} X_i$, $Y = \coprod_{i\in I} Y_i$, be the coproducts (that is, direct sums) of the sets, $p_i : X_i \to X$, $q_i : Y_i \to Y$ be the corresponding coprojections (inclusions) and let $R : X \times Y \times M \to L$ be defined by $R(x, y, \alpha) = \bigvee_i R_i(x_i, y_i, \alpha)$. Then the dilation space (X, Y, R, δ_R) is the coproduct of the family $\{X_i, Y_i, R_i, \delta_{R_i} \mid i \in I\}$.*

Proof To make the proof more visible, we restrict to the case of the co-product of two erosion spaces $(X_1, Y_1, R_1, \delta_{R_1})$, $(X_2, Y_2, R_2, \delta_{R_2})$.

First, we have to prove that $(p_i, q_i) : (X_i, Y_i, R_i, \delta_{R_i}) \to (X, Y, R, \delta_R)$ is a continuous transformation. Without loss of generality, we assume that $i = 1$ and hence we have to prove that $\delta_R(p_1(B)) \geq q_1(\delta_{R_1}(B)$ for every $B \in L^{Y_1 \times Y_2}$. We fix $\bar{x}_1 \in X_1 \subset X$. Then $\delta_R(p_1(B))(\bar{x}_1, \alpha) = \bigvee_{y\in Y}(R((\bar{x}_1, x_2), y, \alpha) * p_1(B)(y) \geq \bigvee_{y_1\in Y} R_1(\bar{x}_1, y_1, \alpha) * \bigvee_{y\in Y} B(y) = \bigvee_{y_1\in Y_1} R_1(\bar{x}_1, y_1, \alpha) * B(y_1) = q_1(\delta_{R_1}(B)(\bar{x}_1)$.

Further, let $(Z, U, R_0, \varepsilon_{R_0})$ be a dilation space and let $(\varphi_i, \psi_i) : (X_i, Y_i, R_i, \delta_{R_i} \to (Z, U, R_0, \delta_{R_0})$ be continuous transformations for i=1,2. We dwfinw $\varphi = (\varphi_1, \varphi_2) : X_1 \coprod X_2 \to Z$ and $\psi = (\psi_1, \psi_2) : Y_1 \coprod Y_2 \to Y$ by setting $\varphi(x) = \varphi_i(x)$ iff $x \in X_i$ and $\psi(y) = \psi_i(y)$ iff $y \in Y_i$. Then from the construction it is easy to see that $(\varphi, \psi) : (X, Y, R, \delta_R) \to (Z, U, R_0, \varepsilon_{R_0})$ is a continuous transformation and $(\varphi, \psi) \circ (p_i, q_i) = (\varphi_i, \psi_i) : (X_i, Y_i, R_i, \delta_{R_i}) \to (Z, U, R_0, \delta_{R_0})$ for $i = 1, 2$. □

By Freyd theorem (see e.g. [15, 18.22]) we have the following corollary:

Corollary 4 *Products exist in category $\mathbf{\Delta}$.*

5.3 Categories $\mathbf{E\Delta^{-1}}$ *and* $\mathbf{E^{-1}\Delta}$

Given a many-level L-fuzzy relation $R : X \times Y \times M \to L$, we can interpret $\varepsilon_R : L^X \to L^{Y\times M}$ as the process of transferring L-fuzzy subsets of X to the L-fuzzy subsets of Y at the moment, or at the level $\alpha \in M$. To say it differently, $\varepsilon_R(A)(\cdot, \alpha) \in$

L^Y is a certain cast of the L-fuzzy set A in L^Y at the moment α. The properties (1)–(3) of Proposition 3 correspond to the properties of a stratified Alexandroff fuzzy interior operator (see e.g. [17, 23, 24]) defined on the L-powerset L^X but taking values in L^Y. Moreover, in case R is upper semicontinuous, properties (4), (5) from Proposition 3 allow to conclude that the operator $\varepsilon_R : L^X \to L^{Y\times M}$ behaves as an *M-valued* Alexandroff stratified fuzzy interior operator (see e.g. [5, 12, 17, 24], cf also [2]), that is $\varepsilon_R(A)(\cdot, \bigvee_x \alpha_i) = \bigwedge_i \varepsilon_R(A)(\cdot, \alpha_i)$.

Let now the relation $R^{-1} : Y \times X \times M \to L$ be defined by $R^{-1}(y, x, \alpha) = R(x, y, \alpha)$ and interpreted as the relation *from Y to X*. This enables to interpret $\delta_{R^{-1}} : L^X \to L^{Y\times M}$ as a process of transfer of L-fuzzy subsets of X to the L-fuzzy subsets of Y at the moment, or at the level $\alpha \in M$. To say it differently, $\delta_{R^{-1}}(A)(\cdot, \alpha) \in L^Y$ is a certain cast of the L-fuzzy set A in L^Y. The properties (1)–(3) of Proposition 4 and reformulated for $\delta_{R^{-1}}$ correspond to the properties of a stratified Alexandroff fuzzy closure operator (see e.g. [11, 17, 23, 24]) defined on the L-powerset L^X but taking values in $L^{Y\times}$. Moreover, in case R is upper semicontinuous, property (5) from Proposition 4 means that the operator $\delta_R^{-1} : L^X \to L^{Y\times M}$ behaves as an *M-valued* Alexandroff stratified fuzzy closure operator (see e.g. [5, 8, 12, 24], cf also [2]), that is $\delta_{R^{-1}}(A)(\cdot, \bigvee_i \alpha_i) = \bigvee_i \delta_{R^{-1}}(A)(\cdot, \alpha_i)$. Note that R is upper semicontinuous if and only if R^{-1} is upper semicontinuous.

Referring to Proposition 1, we know that if R is strongly right connected, then $R^{\Rightarrow}(A) \leq R^{\rightarrow}(A)$, and hence, in case R is left and right connected, $\varepsilon_R(A) \leq \delta_{R^{-1}}(A)$ (notice that right connectedness of R is equivalent to the left connectedness of R^{-1}) These observations justify our interpretation of $\varepsilon_R(A)$ as the L-fuzzy interior of $A \in L^X$ transferred to $L^{Y\times M}$ and $\delta_{R^{-1}}(A)$ as the L-fuzzy closure of $A \in L^X$ transferred to $L^{Y\times M}$.

In its turn, the above observations lead us to the idea to view the pair $(\varepsilon_R, \delta_{R^{-1}})$ as a certain topological-type structure induced by relation R from L^X to $L^{Y\times M}$ and to justify the following definition where L and M, as before, are assumed to be fixed.

Definition 14 The tuple $(X, Y, R, \varepsilon_R, \delta_{R^{-1}})$ is called a many-level fuzzy topological $(\varepsilon, \delta^{-1})$-morphology space.

In the definition of a $(\varepsilon, \delta^{-1})$-space no restrictions on the relation $R : X \times Y \times M \to L$ are made. However, as we see from the above observations, the most "smooth" theory will be possible if we assume that R is right and left connected (in this case we have $\varepsilon_R(A) \leq \delta_{R^{-1}}(A)$, and if we assume that R is upper semicontinuous—in this case operators ε_R and $\delta_{R^{-1}}$ are "continuous" along the level scale M, and this property is essential for interpreting operators ε_R and $\delta_{R^{-1}}$ from many-valued topological point of view.

Changing the roles of the sets X and Y, we come to the concept of a*many-level fuzzy topological* $(\varepsilon^{-1}, \delta)$*-morphology space*, in a certain sense dual to the many-level fuzzy topological $(\varepsilon, \delta^{-1})$-morphology space considered above. Since $(R^{-1})^{-1} = R$, all properties of the pair $(\varepsilon_R, \delta_{R^{-1}})$ can be reformulated for the pair $(\varepsilon_{R^{-1}}, \delta_R)$.

The restricted volume of this paper does not allow us to discuss the properties of appropriately defined categories $\mathbf{E}\Delta^{-1}$ and $\mathbf{E}^{-1}\Delta$ of $(X, Y, R, \varepsilon_R, \delta_{R^{-1}})$- and $(X, Y, R, \varepsilon_{R^{-1}}, \delta_R)$ spaces respectively. This work will be continued in future papers.

6 Conclusion

In this paper, we have introduced many-level versions of L-fuzzy relational erosion and dilation operators and initiated their investigations in the framework of the category theory. As the main directions for the further research of many-level fuzzy morphological spaces we view the following.

- In this paper, we just initiated to study categories of many-level fuzzy morphological spaces $\mathbf{E}$ and $\mathbf{\Delta}$ and sketched the definitions of the categories $\mathbf{E\Delta^{-1}}$ and $\mathbf{E^{-1}\Delta}$. The restricted volume of the paper did not permit us to go deeper into this subject. In future, we plan to develop a "full value" categorical approach to many-level fuzzy morphological spaces, in particular to investigate initial and final structures in these categories, limits of diagrams etc. Besides, as an essential part of the future research we view the study of the relation of these categories with some other related categories, in particular the category $\mathbf{RelSet^2}$ used here.
- Here we have only sketched a possible topological interpretation of many-level fuzzy morphological spaces. In future, we plan to study the relations between many-level fuzzy morphological spaces and their many-valued fuzzy topological "remote counterparts" in detail. In particular, we will show that these categories are topological over the category $\mathbf{RelSet^2}$.
- Probably, quite interesting, especially from the point of possible application, will be to view many-level fuzzy morphological spaces as certain "remote" fuzzy rough approximation systems. (cf [9, 10, 14, 26], etc.). In particular, it could be useful in the study of big volumes of transformed data.

References

1. Adámek, J., Herrlich, H., Strecker, G.E.: Abstract and Concrete Categories. Wiley, New York (1990)
2. Alexandroff, P.: Diskrete Räume. Mat. Sbornik **2**, 501–518 (1937)
3. Bloch, I.: Duality vs. adjunction for fuzzy mathematical morhology. Fuzzy Sets Syst. **160**, 1858–1867 (2009)
4. Bloch, I., Maitre, H.: Fuzzy mathematical morphology: a comparative study. Pattern Recogn. **28**, 1341–1387 (1995)
5. Brown, L.M., Šostak, A.: Categories of fuzzy topologies in the context of graded ditopologies. Iran. J. Fuzzy Syst. Syst. **11**(6), 1–20 (2014)
6. De Baets, B., Kerre, E.E., Gupta, M.: The fundamentals of fuzzy mathematical morphology Part I: basic concepts. Int. J. Gen. Syst. **23**, 155–171 (1995)
7. De Baets, B., De, Kerre, E.E., Gupta, M.: The fundamentals of fuzzy mathematical morphology Part II: idempotence, convexity and decomposition. Int. J. Gen. Syst. **23**, 307–322 (1995)
8. Chen, P., Zhang, D.: Alexandroff L-cotopological spaces. Fuzzy Sets Syst. **161**, 2505–2525 (2010)
9. Dubois, D., Prade, H.: Rough fuzzy sets and fuzzy rough sets. Int. J. Gen. Syst. **17**, 191–209 (1990)
10. Eļkins, A., Šostak, A., Uļjane, I.: On a category of extensional fuzzy rough approximation operators. In: Communication in Computer Information Science, vol. 611 (2016)

11. Fang, J.-M.: I-fuzzy Alexandrov topologies and specialization orders. Fuzzy Sets Syst. **158**, 2359–2374 (2007)
12. Han, S.-E., Šostak, A.: On the measure of M-rough approximation of L-fuzzy sets. Soft Comput. **22**, 2843–2855 (2018)
13. Höhle, U.: M-valued sets and sheaves over integral commutative cl-monoids, Chapter 2. In: Rodabaugh, S.E., Klement, E.P., Höhle, U. (eds.) Applications of Category Theory to Fuzzy Sets, pp. 33–73. Kluwer Academic Publishers (1992)
14. Pawlak, Z.: Rough sets. Int. J. Comput. Inf. Sci. **11**, 341–356 (1982)
15. Herrlich, H., Strecker, G.E.: Category Theory. Heldermann Verlag, Berlin (1979)
16. Leinster, T.: Basic Category Theory. Cambridge University Press, Cambridge (2014)
17. Liu, Y., Luo, M.: Fuzzy Topology Advances in Fuzzy Systems - Applications and Topology. World Scientific. Singapore, New Jersey, London, Hong Kong (1997)
18. Matheron, G.: Random Sets and Integral Geometry. Willy (1975)
19. Madrid, N., Ojeda-Aciego, M., Medina, J., Perfilieva, I.: L-fuzzy relational mathematical morphology based on adjoint triples. Inf. Sci. **474**, 75–89 (2019)
20. Serra, J.: Image Analysis and Mathematical Morphology, I. Academic Press (1982)
21. Šostak, A., Uļjane, I.: M-bornologies on L-valued sets. Adv. Intell. Syst. Comput. **643**, 450–462
22. Šostak, A., Uļjane, I.: Bornological structures on many-valued sets. RAD HAZU Matematiçke Znanosti **21**, 145–170 (2017)
23. Šostak, A.: Two decades of fuzzy topology: basic ideas, notions and results. Russ. Math. Surv. **44**, 125–186 (1989)
24. Šostak, A.: Basic structures of fuzzy topology. J. Math. Sci. **78**, 662–701 (1996)
25. Šostak, A., Uļjane, I., Eļkins, A: On the measure of many-level fuzzy rough approximation for L-fuzzy sets. Stud. Comput. Intell. **819**, 183–190 (2020)
26. Yao, Y., She, Y.: Rough set models in multigranulationl spaces. Inf. Sci. **327**, 40–56 (2016)

Congruences on Lattices and Lattice-Valued Functions

Branimir Šešelja and Andreja Tepavčević

Abstract For a complete lattice L and an L-valued function μ on a domain X, the cuts of μ determine a residuated map f from L to the power set of X ordered dually to inclusion. We describe a class of complete lattices for which the kernel of f is a complete congruence on L. Conversely, every complete congruence on a complete lattice L is uniquely determined by a suitable L-valued function μ on an arbitrary domain, as the kernel of a residuated map which sends every element $p \in L$ into the corresponding cut μ_p. As an application, using residuated maps we get a representation of finite lattices by meet-irreducible elements.

Keywords L-valued functions · Residuated maps · Closure operators · Cuts

1 Introduction

Lattice-valued functions appeared within the fuzzy set theory under the name of L-fuzzy sets, see the first relevant paper [12] by Goguen. Then the topic became widely investigated, one of the first book dealing with lattice-valued functions was [20] by Negoita and Ralescu. Over the years there were attempts to choose an appropriate lattice which would serve as a suitable order and algebraic structure for dealing with

Supported by Ministry of Education, Science and Technological Development, Republic of Serbia, Grant No. 174013.

B. Šešelja · A. Tepavčević
Faculty of Sciences, Department of Mathematics and Informatics, University of Novi Sad, Trg Dositeja Obradovića 4, 21000 Novi Sad, Serbia
e-mail: seselja@dmi.uns.ac.rs

A. Tepavčević (✉)
Mathematical Institute of the Serbian Academy of Sciences and Arts, Kneza Mihaila 36, 11001 Belgrade, Serbia
e-mail: andreja@dmi.uns.ac.rs

M. E. Cornejo et al. (eds.), *Computational Intelligence and Mathematics for Tackling Complex Problems 2*, Studies in Computational Intelligence 955,
https://doi.org/10.1007/978-3-030-88817-6_25

the fuzzy logic. It turned out that residuated lattices and related ordered structures fulfil this task. Still, the basic complete lattices remained an obligatory tool for investigating general properties of lattice-valued (fuzzy) structures. Presently there are many expository books dealing with lattice-valued structures, maybe the best known is the one by Bělohlávek, [3], mostly oriented to residuated lattices. We also mention the overview articles [23, 24] by the present authors. If L is a complete lattice with no additional operations, then in the field of L-valued functions, p-cuts or cuts are one of the basic tools. In papers [14–16] cuts are used in investigations of lattice-valued Boolean functions and in [18], cuts determine particular up-sets related to lattice-valued maps. In the present research the role of cuts is essential, therefore we deal with classical complete lattices. Basic properties of cuts are formulated by the meet operation in L, while in residuated lattices, in analogue context, additional operation—multiplication is used. Under such conditions cuts loose some important properties.

The second tool that we use in this investigation are residuated maps, closely related to Galois connections. They are used in order theory, acting analogously to homomorphisms in the field of algebraic structures; they are isotone and compatible with arbitrary joins. For detailed presentations we refer to books [5] by Blyth and Janowitz, [4] by Blyth, [13] by Grätzer and [8] by Caspard at al.

Up to our knowledge, in the literature connections among the above topics—lattice-valued functions and residuated and related maps exist in the form of Galois connections or adjunctions. Galois functions were generalized to the fuzzy case by Bělohlávek in [2]. Then also in [6, 7, 10], Cabrera et al. investigate Galois connections in the framework of fuzzy preordered structures, using particular fuzzy equivalence relations; the membership-values structure is a residuated lattice.

Here we apply lattice-valued functions (fuzzy sets) to characterize congruences on complete lattices. We use the well known connections among congruences and homomorphisms.

Starting with a complete lattice L and a particular L-valued function μ on a suitably chosen domain X, we show that the cuts of μ determine a residuated mapping f from L to the power set of X ordered dually to inclusion. We identify a wide class of complete lattices for which the kernel of f is a complete congruence on L. In particular, if the set of values of μ is a meet-dense subset of a complete lattice L, then the kernel of f is the diagonal of L.

Conversely, we prove that for every complete congruence $\sim$ on a complete lattice L there is a nonempty subset M of L and a map $\mu : L \to \mathcal{P}(M)$, so that $\sim$ is the kernel of a residuated map determined by the cuts of μ.

As an application, we show that for finite lattices, the mentioned residuated maps can naturally be used for representing these lattices by the posets of their meet-irreducible elements. For our previous research of the similar problems related to finite lattices, see [11, 22].

2 Preliminaries

We deal with ordered sets and lattices, for the necessary notions and notations see e.g., [9].

For posets P and Q, a map $f : P \to Q$ is **residuated** if there exists a map $g : Q \to P$ such that for all $x \in P$ and $y \in Q$

$$f(x) \leq y \text{ is equivalent to } x \leq g(y). \tag{1}$$

The map g is uniquely determined by f and it is called the **residual** of f ([4, 13]). In addition, both mappings are isotone.

The following characterizations can be found in [4, 8, 13].

Proposition 1 *A map $f : P \to Q$ is residuated if and only if any of the following statements holds:*

(a) for every $q \in Q$ there exists

$$\max\{x \in P \mid f(x) \leq y\} = \bigvee(x \in P \mid f(x) \leq y) = g(y); \tag{2}$$

(b) for every $q \in Q$, $f^{-1}(\downarrow q)$ is a principal ideal (principal down-set) in P;

(c) f is isotone and there exists an isotone map g from Q to P such that $I_P \leq f \circ g$ and $g \circ f \leq I_Q$.

Proposition 2 *A map $f : H \to L$, where H and L are complete lattices, is residuated if and only if any of the following statements hold:*

(i) f is completely join preserving. Then the residual is completely meet preserving.

(ii) f is join preserving and $f(0_H) = 0_L$.

An isotone map $C : P \to P$ is a **closure operator** on a poset $(P, \leq)$ if $C = C \circ C \geq I_P$. C is a **dual** closure operator on $(P, \leq)$ if $C = C \circ C \leq I_P$.

If $p = C(p)$, then p is a **closed element** under C.

As usual, if a closure C is a unary operation on the power set $\mathcal{P}(X)$ of a set X, then C is said to be a closure operator **on** X.

Lemma 3 *Let C be a closure (dual closure) operator on a lattice L. Then,*

1. *The subset of all closed elements of L is closed under meets (joins) in L.*
2. *The top (bottom) element of L is closed under C.*

Lemma 4 *The kernel $\sim$ of a closure (dual closure) operator C on a complete lattice L fulfils:*

1. *Each equivalence class of $\sim$ possesses the top (bottom) element which is closed under C.*
2. *The set $L/\sim$ (denoted also by L/C) can be ordered: $[x]_\sim \leq [y]_\sim$ iff $C(x) \leq C(y)$ in L.*

3. *The poset* $(L/\sim, \leq)$ *is a lattice isomorphic with the lattice of closed elements of* L*, under* C.

Proposition 5 *If* P *is an ordered set then* $C : P \to P$ *is a closure mapping if and only if there is an ordered set* Q *and a residuated mapping* $f : P \to Q$ *such that* $C = g \circ f$*, where* g *is the residual of* f.

Recall that a **closure system** over a nonempty set X is a collection of subsets of X closed under all set intersections over X (including thus X).

Lattice-valued or **L-valued functions** are here considered to be mappings from a non-empty set X (domain) into a complete lattice L (co-domain).

Let L be a complete lattice and $\mu : X \to L$ an L-valued function on a set X. For $p \in L$, a p**-cut**, or a **cut** of μ is a subset μ_p of X defined by

$$\mu_p = \{x \in X \mid \mu(x) \geq p\} = \mu^{-1}(\uparrow p).$$

The basic properties of cuts of lattice-valued functions are the following.

Proposition 6 *Let* L *be a complete lattice and* $\mu : X \to L$ *an* L*-valued function on a set* X*. Then*

$$\textit{for}\ \ p, q \in L, \ \ p \leq q \quad \textit{implies} \quad \mu_q \subseteq \mu_p\,; \tag{3}$$

$$\bigcap(\mu_p \mid p \in M \subseteq L) = \mu_{\bigvee p}\,; \tag{4}$$

$$\textit{for every}\ x \in X, \quad \mu(x) = \bigvee(p \in L \mid x \in \mu_p). \tag{5}$$

We denote by μ_L the collection of cuts of $\mu : X \to L$: $\mu_L := \{\mu_p \mid p \in L\}$.

As straightforward consequences of formulas (4) and (5) we have:

Proposition 7 *The collection* μ_L *of cuts of a lattice-valued function* μ *from* X *to* L *is a closure system over* X.

Proposition 8 ([17]) *Let* $\mu : X \to L$ *be an* L*-valued function on* X *and let the power set* $\mathcal{P}(X)$ *be ordered dually to inclusion. Consider the functions*

$$f : L \to \mathcal{P}(X), \quad f(p) := \mu_p, \quad \textit{and} \tag{6}$$

$$g : \mathcal{P}(X) \to L, \quad g(Y) := \bigvee(p \in L \mid \mu_p \supseteq Y). \tag{7}$$

Then f *is a residuated function and* g *is the corresponding residual.*

We say that the residual map f in Proposition 8 is **induced** by the L-valued function μ.

3 Results

3.1 *Complete Congruences on L Induced by L-Valued Functions*

In this part, for a complete lattice L, we analyze complete congruences on L in the framework of residuated maps induced by L-valued functions.

Let L be a complete lattice and $\mu : X \to L$ be an L-valued function on an arbitrary nonempty domain X. Then the map f from L to the power set $\mathcal{P}(X)$ ordered dually to inclusion, such that $f(p) = \mu_p$, is residuated (f induced by μ, Proposition 3). Let also the relation $\approx_\mu$ on L be the kernel of f:

$$p \approx_\mu q \quad \text{if and only if} \quad \mu_p = \mu_q. \tag{8}$$

By the definition of a cut, it is easy to check that for $p, q \in L$,

$$p \approx_\mu q \ \text{ if and only if } \ \uparrow p \cap \mu(X) = \uparrow q \cap \mu(X). \tag{9}$$

As usual, $\mu(X) = \{p \in L \mid p = \mu(x), \text{ for some } x \in X\}$.

Proposition 9 *Let $\mu : X \to L$ be an L-valued function and $\approx_\mu$ be a relation on L, defined by (8). Then:*

(i) $\approx_\mu$ is a complete congruence relation on the join-semilattice reduct of L.

(ii) the map $p \mapsto \bigvee[p]_{\approx_\mu}$ is a closure operator on L.

The poset $(\{\bigvee[p]_{\approx_\mu} \mid p \in L\}, \leq)$, as a subposet of L, is a complete lattice by itself, but not as a sublattice of L. Further, since by Proposition 9 (i), the relation $\approx_\mu$ is a join-semilattice congruence on the complete lattice L, there is an order $\leq$ on $L/\approx_\mu$ induced by the order $\leq$ on L:

$$[p]_{\approx_\mu} \leq [q]_{\approx_\mu} \ \text{ if and only if } \ \mu_q \subseteq \mu_p, \tag{10}$$

and we get the lattice $(L/\approx_\mu, \leq)$. In addition, we can replace the mapping $f : L \to \mathcal{P}(X)$, by the map φ from L (on)to the lattice $(f(L), \supseteq) = (\mu_L, \supseteq)$ (recall that $\mu_L = \{\mu_p \mid p \in L\}$), so that φ is defined in the same way as f: $\varphi(p) = \mu_p$. The proof of the proposition in the sequel follows by Proposition 9, since f is a join semilattice homomorphism.

We have the following isomorphisms.

Theorem 10 *For $\mu : X \to L$, the following isomorphisms of lattices hold:*

$$(L/\approx_\mu, \leq) \cong (\{\bigvee[p]_{\approx_\mu} \mid p \in L\}, \leq) \cong (\mu_L, \supseteq). \tag{11}$$

Relation $\approx_\mu$ is not generally a congruence on L as a lattice. Here we give conditions under which it is the case.

We say that the a complete lattice L is **dually spatial** if every element of L is a meet of completely meet-irreducible elements (this is the dual notion of a spatial lattice, see [19, 21]). Examples of dually spatial lattices are all finite lattices, then complete, dually atomistic lattices etc. In the case of dually spatial lattices, the subset M of completely meet-irreducible elements is a *meet-dense* in L: every element of L is a meet of a subset of M.

Recall that an element a of a lattice L is *infinitely distributive* if for $\{p_i \mid i \in I\} \subseteq L$, $a \vee \bigwedge_i p_i = \bigwedge(a \vee p_i)$.

Theorem 11 *Let L be a dually spatial lattice and $\mu : X \to L$ an L-valued function on a nonempty domain X, such that $\mu(X)$ consists of some infinitely distributive completely meet-irreducible elements in L. Then $\approx_\mu$ is a complete congruence relation on L.*

Lemma 12 *For $\mu : X \to L$, if $\mu(X)$ is meet-dense in a complete lattice L, then the residuated map $f : L \to \mathcal{P}(X)$ induced by μ is an injection.*

A straightforward consequence of Lemma 12 is:

Corollary 13 *For $\mu : X \to L$, if $\mu(X)$ is meet-dense in a complete lattice L, then the relation $\approx_\mu$ is the smallest congruence relation, diagonal Δ_L, on L.*

One can check that the same result, i.e., that $\approx_\mu$ is a diagonal relation on L, is obtained also in the case when a meet-dense set is a *subset* of $\mu(X)$.

Next we formulate a finite version of the result presented in Theorem 11, together with an application to quotient lattices.

Recall that an element a of a lattice L is *distributive* if for all $x, y \in L$, $a \vee (x \wedge y) = (a \vee x) \wedge (a \vee y)$.

Theorem 14 *Let L be a finite lattice, let M be a subset of L consisting of some distributive meet-irreducible elements, and $\mu : M \to L$, $\mu(m) = m$. Then the following hold:*

(i) $\approx_\mu$ is a congruence relation on L.

(ii) The set $[M] := \{[m]_{\approx_\mu} \mid m \in M\}$, consists of all meet-irreducible elements of the quotient lattice $(L/\approx_\mu)$.

Corollary 15 *Let L be a finite distributive lattice and $M \subseteq L$, consisting of (some) meet-irreducible elements in L. Then $\approx_\mu$ is a congruence relation on L.*

3.2 From Congruences to Residuated Maps

Next we formulate the converse of the result in Sect. 3.1. We show that *every complete congruence on a complete lattice L is the kernel of the residuated map induced by a particular L-valued function whose co-domain is L.*

First we deal with a congruence on the join-semilattice reduct of L. We show that it can be the kernel of a residuated map, as in Proposition 8.

We need the following technical lemma, the proof of which is obvious.

Lemma 16 *Let L be a complete lattice, let M be a nonempty subset of L and $\mu : M \to L$, such that for every $m \in M$, $\mu(m) = m$. Then for every $p \in L$, $\mu_p = \uparrow p \cap M$.*

Theorem 17 *Let $\sim$ be a complete congruence on the join-semilattice reduct of a complete lattice L. Let M be a set of maximum elements in $\sim$-classes:*

$$M = \{m \in L \mid m = \bigvee [p]_\sim \text{ for some } p \in L\}.$$

Let also $\mu : M \to L$ be the L-valued function given by $\mu(m) = m$. Then the function $f : L \to \mathcal{P}(M)$ such that $f(p) = \mu_p$ and $\mathcal{P}(M)$ is ordered dually to inclusion is residuated and $\sim$ is the kernel of f.

Straightforwardly from Theorem 17, we get the answer for congruences.

Corollary 18 *For every complete congruence $\sim$ on a complete lattice L, there is a set M, an L-valued function $\mu : M \to L$ and a residuated map $f : L \to \mathcal{P}(M)$ induced by μ by $f(p) = \mu_p$, so that $\sim$ is the kernel of f.*

3.3 Representation of Finite Lattices

Here we use residuated maps induced by L-valued functions to deal with finite lattices with a given poset of meet-irreducible elements.

In particular, for a finite poset M we denote by L^M the collection of all upsets of M. By Birkhoff's theorem, a finite *distributive* lattice in which M is the poset of meet-irreducibles is isomorphic with $(L^M, \supseteq)$ under $p \mapsto \uparrow p \cap M$. Dealing with residuated functions in this case, we replace the power set of M by L^M.

Proposition 19 *Let L be a finite lattice in which M is the poset of meet-irreducibles and $\mu : M \to L$, $\mu(m) = m$. Then $\approx_\mu$ is Δ_L.*

Proof. Indeed, the map $f : L \to L^M$, given by $f(p) = \uparrow p \cap M$ where L^M is ordered dually to inclusion, is residuated, induced by μ, since $\uparrow p \cap M = \mu_p$. Therefore, $\approx_\mu$ is Δ_L by Theorem 13, since M is clearly meet-dense in L.

Observe that under conditions of Proposition 19, the map $g : L^M \to L$, such that

$$g(N) := \bigvee (p \mid f(p) \supseteq N, N \in L^M) \tag{12}$$

is the residual of function f. We proceed by analyzing this residual.

Proposition 20 *Let L be finite lattice L with the set M of meet-irreducibles and $\mu : M \to L$, $\mu(m) = m$. Then the map $C : L^M \to L^M$ such that $C = g \circ f$, where f and g are induced by μ, as in Proposition 19, is a dual closure operator on $(L^M, \supseteq)$, namely $C(Y) = \mu_{g(Y)}$, $Y \in L^M$.*

Next we have a representation theorem for finite lattices in terms of up-sets and residuated functions. Namely, we prove that *any finite lattice with the poset M of meet-irreducibles, is isomorphic to lattice of closed elements of a dual closure in the distributive lattice L^M ordered by $\supseteq$.*

Theorem 21 *Let L be a finite lattice and M the set of all meet-irreducible elements in L. Let also the function $g : L^M \to L$ be given by (12). Then $L \cong (\{\mu_{g(N)} \mid N \in L^M\}, \supseteq)$.*

By Lemma 4 (the dual), we have also the following.

Corollary 22 *Let L be a finite lattice and M the set of all meet-irreducible elements in L. Then $(L, \leq) \cong (L^M/{\sim_C}, \leq)$, where $\sim_C$ is the kernel of the dual closure operator C on L^M defined in Proposition 20.*

The converse is also valid.

Theorem 23 *Let M be a finite poset and C a dual closure operation on $(L^M, \supseteq)$ fulfilling: for every $m \in M$ the class $[m]_{\sim_C}$ is a singleton, where $\sim_C$ is the kernel of C. Then the poset of meet-irreducibles of the lattice $(L^M/{\sim_C}, \leq)$ is isomorphic with M.*

Example

The lattice L in Fig. 1a possesses five meet-irreducible elements p, q, r, s, t denoted by filled circles, forming a poset M, represented in Fig. 1b. According to Corollary 22, L is isomorphic with the lattice $(L^M/{\sim_C}, \leq)$. Observe that $\sim_C$ is not a congruence on L^M; this quotient is isomorphic with the lattice of closed elements of the dual closure C on $(L^M, \supseteq)$, illustrating Theorem 21. (In Fig. 1c, up-sets of M are denoted as words: e.g, pq instead of $\{p, q\}$ etc.).

Figure 2 uses the same finite lattice L and illustrates Theorem 14. Among the meet-irreducibles of the lattice L, only p and q are distributive, $N = \{p, q\}$. The L-valued function inducing the corresponding residuated function is $\nu : N \to L$, $\nu(x) = x$, and the corresponding congruence $\approx_\nu$ is indicated in Fig. 1c. Recall that for $x, y \in L, x \approx_\nu y$ if and only if $\nu_x = \nu_y$ if and only if $\uparrow x \cap \{p, q\} = \uparrow y \cap \{p, q\}$. Hence the lattice $L/{\approx_\nu}$ is a three-element chain.

Finally, N could be a nonempty subset of $\{p, q\}$, namely we can take $N = \{p\}$, or $N = \{q\}$. In both cases we get a congruence splitting L into two classes (in each case joining two neighboring classes of the congruence in Fig. 2).

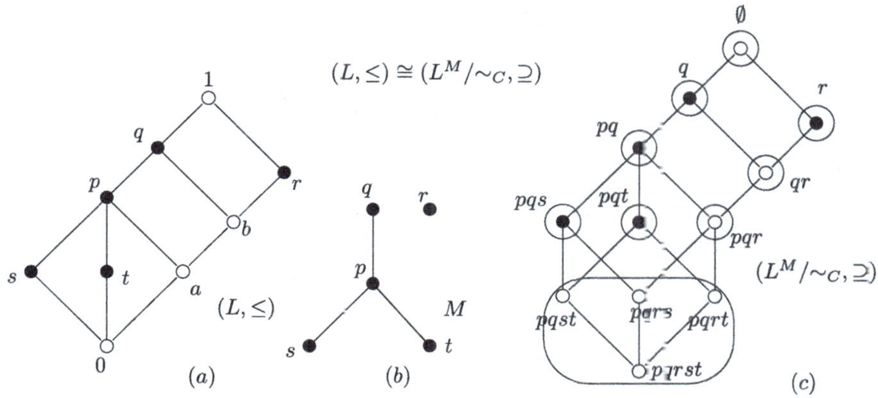

Fig. 1 (**a**) Lattice L; (**b**) The set of all meet-irreducible elements M of L; (**c**) The set of all up-sets on M

Fig. 2 A factor lattice, an illustration of Theorem 14

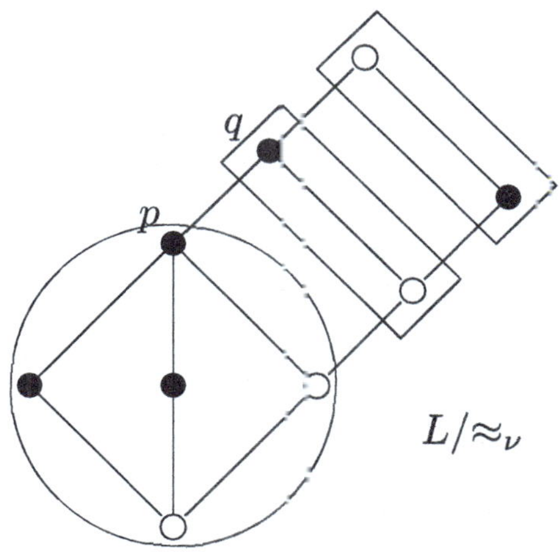

4 Conclusion

Since residuated maps are closely related to Galois connections, our next task is to apply the obtained theoretical results in formal concept analysis. Moreover, the transformation of the L-valued fuzzy set $\mu : X \to L$ into a residuated map f_μ is also used in the theory of soft sets (see, e.g. [1]). In the soft sets' theory, every L-valued fuzzy set μ is transformed into the soft set $f_\mu : L \to 2^X$ by the corresponding construction. Hence in our future investigation we plan use the synergy of the two theories in order to extract new knowledge in the lattice theory and applications.

References

1. Aktas, H., Cagman, N.: Soft sets and soft groups. Inf. Sci. **177**, 2726–2735 (2007)
2. Bělohlávek, R.: Fuzzy Galois connections. Math. Log. Q. **45**(4), 497–504 (1999)
3. Bělohlávek, R.: Fuzzy Relational Systems. Kluwer Academic Publishers, Dordrecht (2002)
4. Blyth, T.S.: Lattices and Ordered Algebraic Structures. Springer (2005)
5. Blyth, T.S., Janowitz, M.F.: Residuation Theory. Elsevier (2014)
6. Cabrera, I.P., Cordero, P., Garca-Pardo, F., Ojeda-Aciego, M., De Baets, B.: On the construction of adjunctions between a fuzzy preposet and an unstructured set. Fuzzy Sets Syst. **320**(1), 81–92 (2017)
7. Cabrera, I.P., Cordero, P., Garca-Pardo, F., Ojeda-Aciego, M., De Baets, B.: Galois connections between a fuzzy preordered structure and a general fuzzy structure. IEEE Trans. Fuzzy Syst. **26**(3), 1274–1287 (2018)
8. Caspard, N., Leclerc, B., Monjardet, B.: Finite Ordered Sets: Concepts, Results and Uses. Cambridge University Press (2012)
9. Davey, B.A., Priestley, H.A.: Introduction to Lattices and Order. Cambridge University Press (1992)
10. Garca-Pardo, F., Cabrera, I.P., Cordero, P., Ojeda-Aciego, M.: On Galois connections and soft computing. In: International Work-Conference on Artificial Neural Networks, pp. 224–235. Springer (2013)
11. Erné, M., Šešelja, B., Tepavčević, A.: Posets generated by irreducible elements. Order **20**(1), 79–89 (2003)
12. Goguen, J.A.: L-fuzzy Sets. J. Math. Anal. Appl. **18**, 145–174 (1967)
13. Grätzer, G.: General Lattice Theory, 2nd edn. Birkhäuser Verlag (2003)
14. Horváth, E.K., Šešelja, B., Tepavčević, A.: Isotone lattice-valued Boolean functions and cuts. Acta Sci. Math. Szeged. **81**, 375–80 (2015)
15. Horváth, E.K., Šešelja, B., Tepavčević, A.: A note on lattice variant of thresholdness of Boolean functions. Miskolc Math. Notes **17**(1), 293–304 (2016)
16. Horváth, E.K., Šešelja, B., Tepavčević, A.: Cut approach to invariance groups of lattice-valued functions. Soft Comput. **21**(4), 853–859 (2017)
17. Horváth, E.K., Radeleczki, S., Šešelja, B., Tepavčević, A.: Cuts of poset-valued functions in the framework of residuated maps, submitted
18. Jiménez, J., Montes, S., Šešelja, B., Tepavčević, A.: On lattice valued up-sets and down-sets. Fuzzy Sets Syst. **151**, 1699–1710 (2010)
19. Johnstone, P.: Stone Spaces, pp. 43 (40). Cambridge University Press, Cambridge (1982)
20. Negoita, C.V., Ralescu, D.A.: Applications of Fuzzy Sets to System Analysis. Birkhäuser Verlag, Basel (1975)
21. Santocanale, L., Wehrung, F.: Varieties of lattices with geometric descriptions. Order **30**(1), 13–38 (2013)
22. Šešelja, B., Tepavčević, A.: Collection of finite lattices generated by a poset. Order **17**(2), 129–139 (2000)
23. Šešelja, B., Tepavčević, A.: Completion of ordered structures by cuts of fuzzy sets an overview. Fuzzy Sets Syst. **136**, 1–19 (2003)
24. Šešelja, B., Tepavčević, A.: Representing ordered structures by fuzzy sets an overview. Fuzzy Sets Syst. **136**, 21–39 (2003)

Some Roughness Features of Fuzzy Sets

Zoltán Ernő Csajbók

Abstract Studying rough calculus was initiated by Z. Pawlak. In this paper, a few roughness features of fuzzy sets will be investigated, looking at the fuzzy membership functions from the point of view of rough continuity and rough approximations of real functions. In the paper, these subfields of rough calculus accurately follow Pawlak's ideas.

Keywords Rough set theory · Rough calculus · Rough continuity · Rough approximation of functions · Fuzzy sets

1 Introduction

In the mid-1990s, Z. Pawlak, based on the rough set theory (RST) [15, 16, 23], initiated the study of rough calculus of real functions in many papers [17, 19–21]. He started the investigation of different subfields of rough calculus such as rough continuity–discontinuity, rough derivatives–integrals, rough differential equations, etc. In [8], the rough continuity–discontinuity is treated in Pawlak's spirit. Other papers follow a different method using rough granular computing [25, 26].

Roughness features of fuzzy membership functions can be investigated via either rough membership functions [22] or rough function approximations [17, 20]. These two approaches are not equivalent. The first one may relate to the measure theory, whereas the second approach has a topological nature [20].

In [9], possible coincidences between rough and fuzzy membership functions were outlined. In this paper, rough continuity–discontinuity and rough function approximations are applied to reveal some roughness features of fuzzy sets.

Section 2 outlines some preliminary notations for the sake of full clarity. It also defines Pawlak approximation space and its fundamental notions. Section 3 surveys

Z. E. Csajbók (✉)
Faculty of Health, Department of Health Informatics, University of Debrecen, Sóstói út 2-4, Nyíregyháza H-4406, Hungary
e-mail: csajbok.zoltan@foh.unideb.hu

M. E. Cornejo et al. (eds.), *Computational Intelligence and Mathematics for Tackling Complex Problems 2*, Studies in Computational Intelligence 955,
https://doi.org/10.1007/978-3-030-88817-6_26

the notion of rough numbers following Pawlak's approach. Section 4, on the one hand, reviews rough continuity–discontinuity mainly based on [8] (Sect. 4.1); on the other hand, it deals with the pointwise rough approximation of real functions (Sect. 4.2). The last subsection contains new results. Section 5 covers the main matter of the paper, namely, some roughness interpretations of real fuzzy membership functions relaying on their rough continuity–discontinuity and pointwise rough approximations.

2 Preliminary Notions

If $a, b \in \mathbb{R}$ $(a \leq b)$, $[a, b] = \{x \in \mathbb{R} \mid a \leq x \leq b\}$ and $]a, b[= \{x \in \mathbb{R} \mid a < x < b\}$ denote *closed* and *open* intervals. $[a, a] = \{a\}$ is identified with $a \in \mathbb{R}$. Similarly, it is easy to interpret the open-closed $]a, b]$ and closed-open $[a, b[$ intervals.

In addition, let $[n] = \{0, 1, \ldots, n\} \subseteq \mathbb{N}$ denote a finite set of natural numbers, and let $]n] = \{1, 2, \ldots, n\}$, $[n[= \{0, 1, \ldots, n-1\}$.

If U and V are two nonempty sets, V^U denotes the set of functions with domain U and co-domain V. A function f from V^U is denoted by $f : U \to V, u \mapsto f(u)$, where $u \mapsto f(u)$ is the assignment or mapping rule of f.

Let U be a nonempty set. Functions in $\{0, 1\}^U$ are called *characteristic functions* on U. A *subset* S of U is $S = \chi_S$, where $\chi_S \in \{0, 1\}^U$. A subset of U is also referred to as *classical set*, *crisp set* or simply *set* on U [11, 12].

The *power set* of U is denoted by $\mathcal{P}(U)$.

In the fuzzy set theory context [13, 27, 28], functions in $[0, 1]^U$ are called *membership functions* on U. A *fuzzy set* (FS) F on U is $F = \mu_F$, where $\mu_F \in [0, 1]^U$.

$\mathcal{F}(U)$ denotes the family of all fuzzy sets on U.

Let U be a nonempty set. $\mathsf{PAS}(U) = \langle U, \mathcal{B}, \mathcal{D}_\mathcal{B}, \mathsf{l}, \mathsf{u}\rangle$ is called a *Pawlak approximation space* [7], if

- $\mathcal{B} = \Pi(U)$, i.e., $\mathcal{B}$ is a partition of U, equivalence classes are called *base sets*;
- $\mathcal{D}_\mathcal{B}$ is the set *definable sets* which is defined with the following inductive definition: $\emptyset \in \mathcal{D}_\mathcal{B}$, $\mathcal{B} \subseteq \mathcal{D}_\mathcal{B}$; if $D_1, D_2 \in \mathcal{D}_\mathcal{B}$, then $D_1 \cup D_2 \in \mathcal{D}_\mathcal{B}$.
- *Pawlak lower and upper approximation operator pair* $\langle \mathsf{l}, \mathsf{u}\rangle$ is defined as
  - $\mathsf{l}(S) = \bigcup\{B \in \mathcal{B} \mid B \subseteq S\}$ $(S \in \mathcal{P}(U))$;
  - $\mathsf{u}(S) = \bigcup\{B \in \mathcal{B} \mid B \cap S \neq \emptyset\}$ $(S \in \mathcal{P}(U))$.

In order to identify and characterize the features of Pawlak approximation spaces, the following basic notions are defined. For any $S \in \mathcal{P}(U)$,

- the *boundary* of S is $\mathsf{bnd}(S) = \mathsf{u}(S) \setminus \mathsf{l}(S)$;
- S is *crisp* (*exact*), if $\mathsf{l}(S) = \mathsf{u}(S)$, i.e., $\mathsf{bnd}(S) = \emptyset$;
- S is *rough* (*inexact*), if it is not exact, i.e., $\mathsf{bnd}(S) \neq \emptyset$;
- S is *definable*, if $\mathsf{l}(S) = S = \mathsf{u}(S)$.

3 Rough Numbers

The notion of rough numbers included in this section can be found mainly in Pawlak's papers [17–20].

Let I denote an interval $I = [0, a]$, where $a \in \mathbb{R}^+$. A *categorization* or *discretization* of I is a sequence $S_I = \{x_i\}_{i \in [n]} \subset \mathbb{R} \geq 0$, where $n \geq 1$ and $0 = x_0 < x_1 < \cdots < x_n = a$.

The categorization S_I generates an equivalence relation I_S on I which is defined as follows. If $x, y \in I$, $x I_S y$ if either $x = y = x_i \in S_I$ for some $i \in [n]$ or $x, y \in]x_i, x_{i+1}[$ for some $i \in [n[$. Hence, the partition I/I_S associated with the equivalence relation I_S is:

$$I/I_S = \{\{x_0\},]x_0, x_1[, \{x_1\}, \ldots, \{x_{n-1}\},]x_{n-1}, x_n[, \{x_n\}\},$$

where $[x_i, x_i] = \{x_i\}$ ($i \in [n]$).

Remark 1 In classical analysis, the phrase "partition of I" is used in a slightly different sense. Two compact real intervals are *disjoint* if their intersection is empty. They are *nonoverlapping* if either they are disjoint or their intersection contains at most one point, which is necessarily an endpoint of both intervals ([5], p. 4). Thus, in classical analysis context, a *partition* or *division* of an interval I is a collection of nonoverlapping closed intervals whose union is I ([4], p. 149). □

The block of the partition I/I_S containing $x \in I$ is denoted by $[x]_{I_S}$. If $x \in [x]_{I_S} =]x_i, x_{i+1}[$, $\overline{[x]}_{I_S} = [x_i, x_{i+1}]$. Of course, when $x \in S_I$, $[x]_{I_S} = \overline{[x]}_{I_S} = \{x\}$. So, any $x \in S_I$ is called the *roughly isolated point* in I.

$\mathsf{PAS}(I) = (I, I/I_S, \mathcal{D}_{I/I_S}, \mathsf{l}_S, \mathsf{u}_S)$ is a Pawlak approximation space. In the following, however, the closed intervals of the form $[0, x]$ ($x \in I$) will only be approximated. Therefore,

- $\mathsf{l}_S([0, x]) = \{y \in I \mid [y]_{I_S} \subseteq [0, x]\} = \cup\{[y]_{I_S} \in I/I_S \mid [y]_{I_S} \subseteq [0, x]\}$,
- $\mathsf{u}_S([0, x]) = \{y \in I \mid [y]_{I_S} \cap [0, x] \neq \emptyset\} = \cup\{[y]_{I_S} \in I/I_S \mid [y]_{I_S} \cap [0, x] \neq \emptyset\}$.

Let $\mathsf{l}_S(x) = \max\{y \in S_I \mid y \leq x\}$ and $\mathsf{u}_S(x) = \min\{y \in S_I \mid y \geq x\}$.

Lemma 1 *Let $x \in I$. With the above notations,*

$$\mathsf{l}_S([0, x]) = [0, \mathsf{l}_S(x)], \ \mathsf{u}_S([0, x]) = [0, \mathsf{u}_S(x)], \ \textit{and}\ \mathsf{l}_S(x) \leq x \leq \mathsf{u}_S(x).$$

Thus, $[x]_{I_S} = [\mathsf{l}_S(x), \mathsf{u}_S(x)] = \{x\}$, *if* $x \in S_I$, *and* $[x]_{I_S} =]\mathsf{l}_S(x), \mathsf{u}_S(x)[$, *if* $x \notin S_I$.

Proof It immediately follows from the above definitions. □

It is said that the number $x \in I$ is *exact* with respect to the approximation space $\mathsf{PAS}(I)$, if $\mathsf{l}_S(x) = \mathsf{u}_S(x)$, otherwise x is *inexact* or *rough* [20]. Of course, $x \in I$ is exact if and only if $x \in S_I$. The members of I/I_S are called *rough numbers* with respect to the approximation space $\mathsf{PAS}(I)$.

Pursuant to Lemma 1, any inexact number $x \in I$ can be represented by the interval $[x]_{I_S}$ or, equivalently, by the pair $(\mathsf{l}_S(x), \mathsf{u}_S(x))$ of exact numbers.

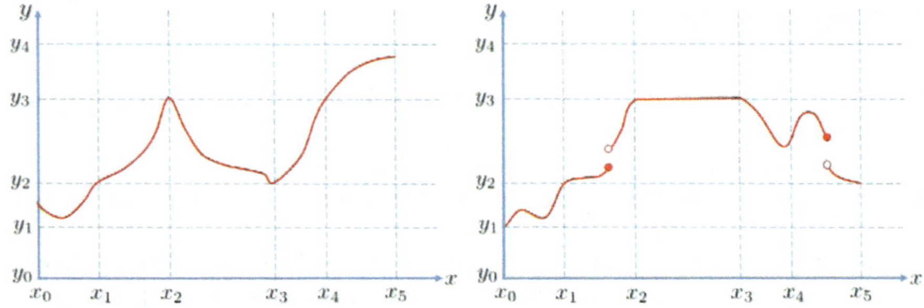

Fig. 1 Roughly continuous real functions

4 Rough Real Functions

Let $I = [0, a_I]$ and $J = [0, a_J]$ be two closed intervals with $a_I, a_J \in \mathbb{R}^+$. Let S_I and P_J be the categorizations of I and J, where $S_I = \{x_i\}_{i \in [n]}$, $P_J = \{y_j\}_{j \in [m]} \subset \mathbb{R} \geq 0$ with $m, n \geq 1$, $0 = x_0 < \ldots < x_n = a_I$ and $0 = y_0 < \ldots < y_m = a_J$. The corresponding Pawlak approximation spaces are $\mathsf{PAS}(I)$ and $\mathsf{PAS}(J)$.

It should be emphasized that all of the considered features of rough real functions occur as the result of the *superposition* of two scales S_I and P_J.

4.1 Rough Continuity–Discontinuity

Let I and J two intervals with categorizations S_I and P_J. Let $f \in J^I$.

Definition 1 ([8, 20], Definition 3) $f \in J^I$ is (S_I, P_J)*–continuous* at x, if $f(\overline{[x]}_{I_S}) \subseteq \overline{[f(x)]}_{J_p}$. Otherwise, f is (S_I, P_J)*–discontinuous* at $x \in I$. □

Proposition 1 ([8], Proposition 1) *A function $f \in J^I$ is (S_I, P_J)–continuous at every $x \in S_I$ roughly isolated point.*

In general, a roughly continuous function neither touches nor intersects any straight line $y = y_j$ ($y_j \in P_J$, $j \in [m]$) on any *proper* subset of every open interval in I/I_S. It should be noted that the adjective "proper" is very important in the previous statement. The converse statement, however, is not true. Figure 1 depicts two roughly continuous real functions. For more details concerning rough continuity–discontinuity, see [8].

Example 1 In the running example, concerning fuzzy sets, let $J = [0, 1]$ with $P_J = \{y_0, y_1, y_2, y_3, y_4\}$, where $y_0 = 0 < y_1 < y_2 < y_3 < y_4 = 1$. Let $I = [0, x_5]$ with $S_I = \{x_0, x_1, x_2, x_3, x_4, x_5\}$, where $x_0 = 0 < x_1 < x_2 < x_3 < x_4 < x_5$. Let $f \in J^I$.

In Fig. 2a, f is roughly continuous, e.g, at x^i, x^{ii}, and x_4, whereas in Fig. 2b, f is roughly discontinuous, e.g., at points x^{iii}, x^{iv}, x^v, and x^{vi}. □

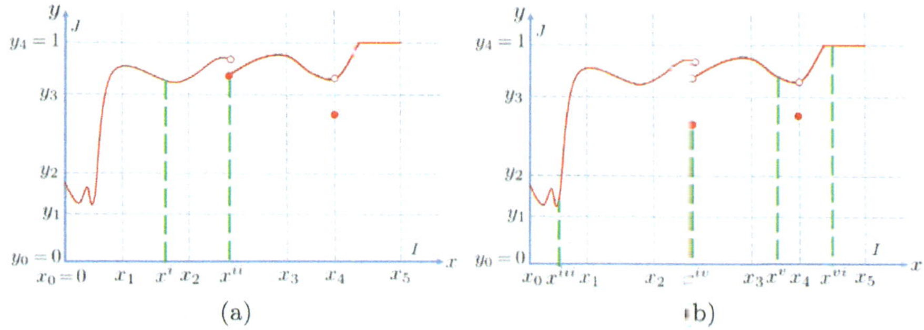

Fig. 2 Rough continuity–discontinuity

4.2 *Pointwise Lower and Upper Approximations of Real Functions*

Definition 2 ([17, 20]) Let $f \in J^I$. The *pointwise* (S_I, P_J)*–lower* and (S_I, P_J)*–upper approximations* of f are the functions $\underline{f}, \overline{f} : I \to P_J$, where

$x \mapsto \underline{f}(x) = \max\{y \in P_J \mid y \leq f(x)\}$, $x \mapsto \overline{f}(x) = \min\{y \in P_J \mid y \geq f(x)\}$.

The function f is *exact* at x, if $\underline{f}(x) = \overline{f}(x)$, otherwise f is *inexact* (*rough*) at x. □

Proposition 2 *Let $f \in J^I$. The function f is exact at $x \in I$ if and only if $f(x) = y_j \in P_J$ for some $j \in [m]$.*

Proof (⇒) On the contrary, let us assume that f is exact at x, but $f(x) \notin P_J$. Then, $f(x) \in]y_j, y_{j+1}[$ for some $j \in [m[$, where $y_j < y_{j+1}$. Hence,

$$\underline{f}(x) = \max\{y \in P_J \mid y \leq f(x)\} = y_j \neq \overline{f}(x) = \min\{y \in P_J \mid y \geq f(x)\} = y_{j+1},$$

which is a contradiction.

(⇐) If $f(x) = y_j \in P_J$ for some $j \in [m]$, then

$$\underline{f}(x) = \max\{y \in P_J \mid y \leq y_j\} = \overline{f}(x) = \min\{y \in P_J \mid y \geq y_j\} = y_j,$$

that is, f is exact at $x \in I$. □

Proposition 2 means that f is exact at a point in I iff in this point f intersects or touches a line segment $y = y_j$, where y_j is a categorization point of J.

Example 2 Figure 3 depicts a rough function $f \in J^I$ (solid line). It also depicts the (S_I, P_J)–lower approximation of f (dashed line), and the (S_I, P_J)–upper approximation

f is exact at points $x^i, x^{ii}, x_2, x^{iii}$ and rough at all other points. □

5 Fuzzy Sets from the Point of View of Rough Calculus

Point A. In the mathematical sense, all real functions in $[0, 1]^I$ are fuzzy sets.

> Although this statement is true in a formal mathematical sense, many functions that qualify on the basis of this definition cannot be suitable fuzzy sets. But they *become* fuzzy sets when, and only when, they match some intuitively plausible semantic description of imprecise properties of the objects in X (*X is the universe of objects – the author*). ([6] which is cited in [24], p. 16)

A scheme (S_I, P_J) may serve as a framework to manage the semantic description of membership functions.

Point B. It is well-known the *vertical view* and *horizontal view* of fuzzy sets. According to the latter view, fuzzy sets can uniquely be represented with a family of classical sets which are called *α–level sets* or *α–cuts* [14]. This view is frequently used for the internal representation of fuzzy sets in computers. In this case, first of all, the membership degrees must be discretized (vertical discretization). Of course, the domain can/must also be discretized depending on the considered problem (horizontal discretization) [14].

Unfortunately, there are no general rules for such discretizations. An (S_I, P_J)–roughness categorization of the membership degree and the domain may serve as a formal background for the discretizations in both directions.

Point C. The *core* and *support* of a fuzzy set F are defined by

$$\mathtt{core}(F) = \{u \in U \mid \mu_F(u) = 1\} \text{ and } \mathtt{supp}(F) = \{u \in U \mid \mu_F(u) > 0\}.$$

It is straightforward that they are crisp sets and bound the fuzzy set F: $\chi_{\mathtt{core}(F)} \le \mu_F \le \chi_{\mathtt{supp}(F)}$. However, the functions $\chi_{\mathtt{core}(F)}$ and $\chi_{\mathtt{supp}(F)}$ are rather rough approximations of F. For instance, in Fig. 3, $\chi_{\mathtt{core}(f)} = 0$ and $\chi_{\mathtt{supp}(f)} = 1$ which is insignificant approximations of f. On the contrary, $\underline{f}$ and $\overline{f}$ are much better approximations of f, though they are fuzzy sets.

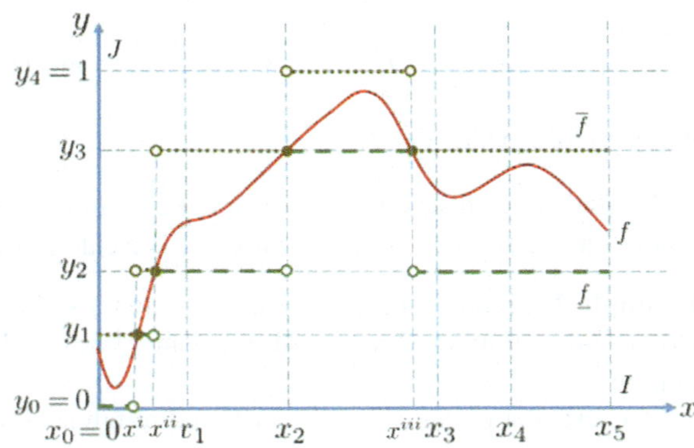

Fig. 3 (S_I, P_J)–lower and (S_I, P_J)–upper approximations of a real function

Point D. Let $D[0, 1] = \{[a, b] \mid 0 \leq a \leq b \leq 1\}$. Of course, $\underline{f}, \overline{f} \in [0, 1]^I$ and $\underline{f} \leq \overline{f}$. Then, the function $f^{IVFS} : I \to D[0, 1]$, $x \mapsto [\underline{f}(x), \overline{f}(x)]$ forms an *interval–valued fuzzy set* [10].

It is well-known (see, e.g., [1]) that every interval–valued fuzzy set $[f, g]$ corresponds to an intuitionistic fuzzy set $(f, 1 - g)$. That is, the function pair $(\underline{f}, 1 - \overline{f})$ forms a *intuitionistic fuzzy set* (IFS) [2, 3]. In this special case, $\underline{f}$ is the IFS *membership function*, $1 - \overline{f}$ is the IFS *nonmembership function*, and $1 - \underline{f} - (1 - \overline{f}) = \overline{f} - \underline{f}$ is the IFS *indeterminacy function*.

6 Conclusion

In this paper, roughness features of fuzzy membership functions have been investigated relying on the notions of rough continuity and rough lower and upper approximations of real functions. This approach, unlike the standard approach, puts certain properties of fuzzy sets in a different light.

Acknowledgements The author would like to thank the anonymous referees for their useful comments and suggestions.

References

1. Atanassov, K., Gargov, G.: Interval valued intuitionistic fuzzy sets. Fuzzy Sets Syst. **31**(3), 343–349 (1989)
2. Atanassov, K.T.: Intuitionistic fuzzy sets. Fuzzy Sets Syst. **20**(1), 87–96 (1986)
3. Atanassov, K.T.: On Intuitionistic Fuzzy Sets Theory, Studies in Fuzziness and Soft Computing, vol. 283. Springer (2012)
4. Bartle, R., Sherbert, D.: Introduction to Real Analysis, 4th edn. Wiley, Incorporated (2011)
5. Bartle, R.G.: A Modern Theory of Integration, Graduate Studies in Mathematics, vol. 32. American Mathematical Society (2001)
6. Bezdek, J.: Editorial: fuzzy models - what are they, and why? IEEE Trans. Fuzzy Syst. **1**, 1–5
7. Ciucci, D., Mihálydeák, T., Csajbók, Z.E.: On exactness, definability and vagueness in partial approximation spaces. Tech. Sci. **18**(3), 203–212 (2015)
8. Csajbók, Z.E.: On the roughly continuous real functions. In: Mihálydeák, T., Min, F., Wang, G., Banerjee, M., Düntsch, I., Suraj, Z., Ciucci, D. (eds.) Rough Sets, pp. 52–65. Springer International Publishing, Cham (2019)
9. Csajbók, Z.E., Ködmön, J.: Roughness and Fuzziness, pp. 23–34. Springer International Publishing, Cham (2020)
10. Deschrijver, G., Kerre, E.E.: On the relationship between some extensions of fuzzy set theory. Fuzzy Sets Syst. **133**(2), 227–235 (2003)
11. Halmos, P.R.: Naive Set Theory. D. Van Nostrand Inc., Princeton (1960)
12. Hayden, S., Zermelo, E., Fraenkel, A., Kennison, J.: Zermelo-Fraenkel Set Theory, Merrill Mathematics Series. C. E. Merrill (1968)
13. Klir, G.J., Yuan, B.: Fuzzy Sets and Fuzzy Logic. Theory and Applications. Prentice Hall, New Jersey (1995)

14. Kruse, R., Borgelt, C., Klawonn, F., Moewes, C., Steinbrecher, M., Held, P.: Computational Intelligence: A Methodological Introduction. Springer, London (2013)
15. Pawlak, Z.: Rough sets. Int. J. Comput. Inf. Sci. **11**(5), 341–356 (1982)
16. Pawlak, Z.: Rough Sets: Theoretical Aspects of Reasoning About Data. Kluwer Academic Publishers, Dordrecht (1991)
17. Pawlak, Z.: Rough real functions. vol. 50. Institute of Computer Science Report. Warsaw University of Technology, Warsaw (1994)
18. Pawlak, Z.: Rough calculus. In: Proceedings of the 2nd Annual Joint Conference on Information Science, pp. 251–271. Wiley, New York (1995)
19. Pawlak, Z.: Rough calculus. vol. 58. Institute of Computer Science Report. Warsaw University of Technology, Warsaw (1995)
20. Pawlak, Z.: Rough sets, rough relations and rough functions. Fundam. Inform. **27**(2/3), 103–108 (1996)
21. Pawlak, Z.: Rough real functions and rough controllers. In: Lin, T., Cercone, N. (eds.) Rough Sets and Data Mining: Analysis of Imprecise Data, pp. 139–147. Kluwer Academic Publishers, Boston (1997)
22. Pawlak, Z., Skowron, A.: Rough membership functions. In: Yager, R.R., Kacprzyk, J., Fedrizzi, M. (eds.) Advances in the Dempster-Shafer Theory of Evidence, pp. 251–271. Wiley, New York (1994)
23. Pawlak, Z., Skowron, A.: Rudiments of rough sets. Inf. Sci. **177**(1), 3–27 (2007)
24. Ross, T.J.: Fuzzy Logic with Engineering Applications, 3rd edn. Wiley (2010)
25. Skowron, A., Stepaniuk, J., Jankowski, A., Bazan, J.G., Swiniarski, R.W.: Rough set based reasoning about changes. Fundam. Inform. **119**(3–4), 421–437 (2012)
26. Szczuka, M.S., Skowron, A., Stepaniuk, J.: Function approximation and quality measures in rough-granular systems. Fundam. Inform. **109**(3), 339–354 (2011)
27. Zadeh, L.A.: Fuzzy sets. Inf. Control **8**(3), 338–353 (1965)
28. Zimmermann, H.J.: Fuzzy Set Theory-and Its Applications, 4th edn. Springer, Netherlands (2001)

Author Index

M. E. Cornejo et al. (eds.), *Computational Intelligence and Mathematics for Tackling Complex Problems 2*, Studies in Computational Intelligence 955,
https://doi.org/10.1007/978-3-030-88817-6

The manufacturer's authorised representative in the EU is Springer Nature Customer Service Centre GmbH, Europaplatz 3, 69115 Heidelberg, Germany. If you have any concerns regarding our products, please contact ProductSafety@springernature.com

Printed and bound by CPI Group (UK) Ltd, Croydon, CR0 4YY
28/09/2026
02224091-0008